Understanding Statistics for the Social Sciences with IBM SPSS

Understanding Statistics for the Social Sciences with IBM SPSS

Robert Ho

CRC Press
Taylor & Francis Group
Boca Raton London New York

CRC Press is an imprint of the
Taylor & Francis Group, an informa business

SPSS was acquired by IBM in October 2009

CRC Press
Taylor & Francis Group
6000 Broken Sound Parkway NW, Suite 300
Boca Raton, FL 33487-2742

Contents

Section II   Inferential Statistics

Preface

This introductory textbook introduces students to basic statistical concepts. It is suitable for undergraduate students in the social sciences where statistics is a core component in their undergraduate program. The book presents clear explanation of basic statistical concepts, the manual calculation of statistical equations, and offers an introduction to the SPSS software program and in particular how to conduct basic statistical analysis using this program via the popular 'point-and-click' method and the 'syntax' method.

In writing this book, I have presented (1) a comprehensive coverage/ explanation of basic statistical concepts, (2) instructions on statistical analysis using the conventional manual procedural steps, and (3) introducing first-year social sciences students to the powerful SPSS software program. A focal point of this book is to show students how easy it is to analyse data using SPSS once they have learned the basics. I believe that learning how to use this very useful and sophisticated statistical software package at the introductory class level has a number of distinct advantages. First, the traditional and conventional cookbook method of instruction simply requires the first-year student to blindly follow a set of procedural steps; while such a technique may yield the correct results, the step-by-step calculations are simply not meaningful to a lot of students and contribute very little to their understanding of the rationale of how the correct results were obtained. Since the main focus of the analysis is on the *interpretation* of the obtained results (and not on the technique used to derive the results) it does not matter how the results were obtained as long as the obtained results are correct. In other words, it would be much more efficient to bypass the conventional cookbook method and learn how to conduct analysis via statistical software packages such as SPSS.

A second advantage of learning the SPSS software package at the introductory class level is that most social sciences students will employ this program in their later years of study. This is because SPSS is one of the most popular of the many statistical packages currently available. Learning how to use this program at the very start of their academic journey not only familiarizes students with the utility of this program but also provides them with the experience to employ the program to conduct complex analyses in their later years of study.

I hope that this book will serve as a useful resource/guide to all students as they begin their journey into the practical world of statistics!

Robert Ho

NOTE: The data sets and SPSS macros employed in the examples in this book can be accessed and downloaded from the following Internet address: http://www.crcpress.com

Author

Robert Ho earned his DPhil from Waikato University, New Zealand, in 1978. He was an associate professor (retired) in the Graduate School of Psychology at the Assumption University of Thailand. His research interests included quantitative methods, health psychology, and social psychology.

Robert passed away in March 2017 following a long battle with cancer. As an academic he was a fantastic Lecturer and Mentor who always had the time to help out with an analysis from the simple to the multivariate. As a long time friend, his loyalty, hospitality, guitar playing, love of music, and appreciation of food will be missed.

1

Introduction to the Scientific Methodology of Research

1.1 Introduction

Some of the common questions many students ask when they begin their research in psychology are, 'Why do we need to do research? After all, what has the conduct of research got to do with the study of human behavior? Can't we simply study human behavior without recourse to understanding the techniques of scientific investigation, without recourse to hypothesis testing', and of course, the mother of all questions, 'without recourse to statistical analysis?' Fair questions! The simple answer is that without objective, empirical, scientific-based verification of social phenomena, science as we know it will simply cease to exist. And without scientific-based empirical research, progress in knowledge, technology, and innovations will stagnate and possibly come to a screeching halt. At the heart of human existence, evolution, growth, and progress are our needs and desire to learn new things, to understand existing phenomena, to be able to explain scientifically why A causes B, and for many social scientists, to be able to understand 'why people behave the way they do'. To be able to do all these things, we need to be well versed in the scientific methodology of research.

1.2 The Scientific Approach versus the Layperson's Approach to Knowledge

A scientific approach to knowledge is different from a layperson's approach to knowledge. A layperson's approach is typically subjective and based on intuition and everyday observations, whereas a scientific approach to knowledge is based on *systematic observation* and *direct experimentation*. For example, a friend of yours tells you that based on his everyday observations he has concluded that men are smarter than women. Surely, as a budding scientist, you have a lot of questions to ask your friend. How many men and women has he met?

Are these men and women representative of their respective populations? And how did he decide how smart they were? It will not be enough for your friend to say to you that he had met lots of men and women and that he is very good at telling how smart people are. You need proof; show me the evidence! The scientific approach to knowledge does not rely on speculation, guesswork, or armchair philosophizing in drawing conclusions about phenomena. It relies on scientific methodology that incorporates (1) the techniques of *random sampling*, (2) choice of *research designs*, and (3) a thorough understanding of *probability theory* and *hypothesis testing*. For all these methodological issues, statistics plays a crucial role in aiding the researcher to draw inferences/conclusions from the data set about some psychological phenomenon.

Before examining the role that statistics plays in scientific research, let's review how these three methodological 'pillars' provide the foundation for the scientific investigation of social/psychological phenomena.

1.2.1 Sampling

Why is the technique of sampling important in research? Sampling is important in social science research because it helps a researcher to generalize results obtained from a specific sample to the population of interest. In conducting research, it is often impossible and not practical to investigate the entire population, unless of course the population is small, like the student population on a university campus. Normally, the research that social scientists conduct is based on very large populations – like the population of Bangkok city, Thailand. To test the entire population of Bangkok residents (~10 million!) will be impossible. As such, research is normally conducted on a sample drawn from the population of interest. However, at the end of the day, it will be necessary for the researcher to be able to generalize the results obtained from the sample back to the population, that is to say, *what is true of the sample must also be true of the population*. To be able to do this, it is critical that the sample selected is a *probability (random) sample* in which all members of a group (population or universe) have an equal and independent chance of being selected into the sample.

1.2.2 Research Designs

When conducting quantitative research, the most common research designs employed can be broadly classified into two types: (a) *between-groups design* and (b) *correlational design*. They differ primarily in terms of the aim of the study and the research questions posed.

1.2.3 Between-Groups Design

When a researcher is interested in investigating whether the manipulation of an independent variable (IV) has some effect on the dependent variable

(DV), then the between-groups design is appropriate. The between-groups design can be further classified into the *univariate* approach and the *multivariate* approach.

1.3 The Univariate Approach

The univariate approach is one in which the research design involves only one DV. For example, if you are interested in investigating whether there is a gender (the IV) difference in problem-solving skills (the DV), and you have a single measure of 'problem-solving skills', then the design is univariate. For example, Table 1.1 shows the problem-solving scores obtained for a sample of 5 male and 5 female subjects.

The univariate design can involve more than one IV and such a design is called a *factorial* design. For example, in addition to investigating gender difference on problem-solving skills, you are also interested in investigating age difference. Suppose that the variable of age is classified into two categories – young/old. Now, the research design involves two IVs (gender and age) and one DV (problem-solving skills). The univariate design will allow for the investigation of the *joint* effect of gender and age on problem-solving skills. More specifically, the univariate design will allow for the 2 × 2 factorial combination of the two variables of gender and age, giving rise to the four groups of 'male-young', 'male-old', 'female-young', 'female-old', and their effects on the DV. For example, Table 1.2 shows the problem-solving scores obtained for a sample of 10 male subjects (5 male-young and 5 male-old) and 10 female subjects (5 female-young and 5 female-old). Regardless of the number of IVs included for testing, the univariate design involves only one DV.

1.3.1 The Multivariate Approach

The multivariate approach is one in which the research design involves more than one DV. For example, apart from investigating whether there is gender

TABLE 1.1

Problem-Solving Scores
as a Function of Gender

Male		Female	
s1	65	s1	76
s2	72	s2	65
s3	59	s3	84
s4	89	s4	68
s5	72	s5	82

TABLE 1.2

Problem-Solving Scores as a
Function of Gender and Age

	Male			Female	
	Young	Old		Young	Old
s1	65	76	s1	56	65
s2	72	65	s2	68	89
s3	59	84	s3	72	92
s4	89	68	s4	69	73
s5	72	82	s5	70	81

difference in problem-solving skills, the researcher is also interested in finding whether there are gender differences in mathematics skills, English language skills, and overall grade point average (GPA). The design of the study is multivariate in that it involves more than one DV (the four DVs are the scores of problem-solving skills, mathematics skills, English language skills, and overall GPA). The advantage of the multivariate approach is that it takes into account the interrelations among the DVs and analyzes the variables together. For example, Table 1.3 shows the scores obtained for the four DVs of problem-solving skills, mathematics skills, English language skills, and overall GPA as a function of gender (males vs. females).

A major advantage of the multivariate design is that it is often used in *'repeated-measures'* studies. The design is particularly useful in investigating the effectiveness of an intervention strategy in a pre- and post-study. For example, a teacher may be interested in finding out whether a new learning strategy introduced by the school will be effective in improving the students'

TABLE 1.3

Problem-Solving Skills, Mathematics Skills, English Language
Skills, and GPA as a Function of Gender

	Problem-Solving Skills	Maths	English	GPA
Male				
s1	65	84	56	3.2
s2	72	79	64	3.6
s3	59	86	61	3.4
s4	89	93	78	3.8
s5	72	89	64	3.7
Female				
s1	76	64	92	3.5
s2	72	72	89	3.6
s3	59	84	91	3.7
s4	89	71	88	3.6
s5	72	79	95	3.8

TABLE 1.4

Pre- and Post-GPA Scores as a Function of a New Learning
Strategy

	Pre-Strategy (GPA Scores)	New Learning Strategy (X)	Post-Strategy (GPA Scores)
s1	2.8	X	3.4
s2	3.1	X	3.3
s3	2.6	X	3.7
s4	2.9	X	3.5
s5	3.0	X	3.6

GPA. A simple multivariate repeated-measures design will answer this question. For example, Table 1.4 shows that the post-strategy GPA scores for a sample of five students have increased over their pre-strategy scores after experiencing the new learning strategy.

1.4 Correlational Design

The correlational approach to investigation is concerned with (1) finding out whether a relationship exists between two variables and (2) determining the *magnitude* and *direction* of this relationship. Unlike the between-groups design, the correlational design does not involve any manipulation of variables, such as manipulating the gender variable – two groups (males and females) – and to see whether there is any difference in problem-solving ability between these two groups. Rather, the correlational approach is concerned with determining whether a naturally occurring set of scores is related to another naturally occurring set of scores. For example, a developmental psychologist may be interested in the relationship between age and height in children. The psychologist selects a group of children for study, and for each child, he/she records their age in years and their height in inches. The psychologist can then calculate the correlation coefficient between these two variables. The correlation coefficient is a number between −1 and +1 that measures both the *strength* and *direction* of the linear relationship between the two variables.

1.5 Hypothesis Testing and Probability Theory

The primary function of hypothesis testing is to see whether our prediction/hypothesis about some social/psychological phenomenon is supported or not. As mentioned earlier, at the very heart of the scientific methodology is an experiment, and part and parcel of conducting an experiment or a research project is to test the predictions/hypotheses generated from a particular theory or past research findings or from our literature review. In testing a hypothesis, data must be collected, analyzed, and interpreted as to whether or not the findings support the study's hypothesis.

1.5.1 Probability

We employ probability to help us make decisions regarding the tests of our hypotheses, that is, whether to accept or reject our hypotheses. When making a decision as to whether to retain or reject a hypothesis, we have to decide at what chance probability level we should reject the hypothesis. For example, if the calculated probability of a test result occurring by chance is very low, then in all likelihood we will reject chance as an explanation for the obtained result and conclude that the result is most likely due to the manipulation of the IV. On the other hand, if the calculated chance probability of occurrence is very high, then we will most likely retain chance as an explanation for the obtained result and conclude that the result is most likely *not* due to the manipulation of the IV but to chance variation.

1.5.2 Statistics and Scientific Research

As mentioned earlier, the scientific methodology incorporates the techniques of random sampling, choice of research designs, and a thorough understanding of probability theory and hypothesis testing. A common tool that is employed in these methodologies to assist the researcher in drawing inferences/conclusions about his/her research is *statistics*.

1.6 Definition of Statistics

Statistics is the science of collecting and learning from data. It is a branch of mathematics concerned with the collection, classification, analysis, and interpretation of numerical facts, for drawing inferences on the basis of their quantifiable likelihood (probability). Statistics also allows the researcher to interpret grouped data, too large to be intelligible by ordinary observation. The field of statistics is subdivided into *descriptive* statistics and *inferential* statistics.

1.6.1 Descriptive Statistics

Descriptive statistics is the branch of statistics that involves organizing, displaying, and understanding data. More specifically, descriptive statistics involves the analysis of data that helps describe, show, or summarize data in a meaningful way such that, for example, patterns might emerge from the data. Descriptive statistics does not, however, allow the researcher to make conclusions beyond the data analyzed or to reach conclusions regarding any hypotheses the researcher might have made. It is simply a way to describe data.

In presenting research findings, descriptive statistics is very important because if raw data were presented alone, it would be difficult to visualize what the data were showing, especially if there were a lot of it. Descriptive statistics, therefore, enables the researcher to present the data in a more meaningful way, which allows simpler interpretation of the data. For example, if we had the IQ scores from 100 students, we may be interested in the average IQ of those students. We would also be interested in the distribution or spread of the IQ scores. Descriptive statistics allows us to do this. Typically, there are two general types of statistics that are used to describe data:

Measures of central tendency—these are ways of describing the *central position* of a frequency distribution for a group of data. In this case, the frequency distribution is simply the distribution and pattern of IQ scores scored by the 100 students from the lowest to the highest. We can describe this central position using a number of statistics, including the *mean, median,* and *mode.* For example, the mean is simply the average IQ score – the central position of the distribution of the 100 IQ scores.

Measures of spread—these are ways of summarizing a group of data by describing how spread out the scores are. For example, the mean (average) IQ score of our 100 students may be 118. However, not all students will have scored 118 on the IQ test. Rather, their scores will be spread out. Some IQ scores will be lower and others higher than the mean score of 118. Measures of spread help us to summarize how spread out these scores are. A number of statistics are available to describe this spread, including the *range, variance,* and *standard deviation.*

1.6.2 Inferential Statistics

Whereas descriptive statistics is the branch of statistics that involves organizing, displaying, and describing data, inferential statistics is the branch of statistics that involves drawing conclusions about a population based on information contained in a sample taken from that population. With inferential statistics, the researcher is trying to draw conclusions that extend beyond the immediate data. Thus, when conducting research, the researcher employs inferential statistics to serve two major purposes: (1) to infer from the sample data the attitudes/opinions of the population and (2) to aid the researcher in making a decision regarding whether an observed difference between groups (e.g., gender difference in problem-solving ability) is a meaningful/reliable difference or one that might have happened by chance alone. Thus, whereas descriptive statistics is used to simply describe what is going on in a data set, inferential statistics is used to infer from the data set to more general conclusions.

2

Introduction to SPSS

2.1 Learning How to Use the SPSS Software Program

When conducting quantitative research, data must be collected and analyzed in order to answer the research questions posed and/or to test the hypotheses posited by the researcher. By convention, students enrolled in introductory social sciences courses (e.g., psychology) have access to introductory statistics textbooks that provide step-by-step instructions on how to carry out such analyses. Such techniques follow a 'cookbook' method in which students are instructed to calculate mathematical functions in a step-by-step procedure that (hopefully!) culminates in the statistical test results that can be used to answer the research questions posed. Unfortunately though, such a cookbook method teaches the students very little about the meaning of the mathematical functions calculated or how these are applied in deriving the statistical test results. In other words, the cookbook method championed by many introductory statistics textbooks may afford the novice students step-by-step instructions on how to calculate the statistics results, but offer little or no understanding of *what* the mathematical functions mean and *why* they are calculated. The end result of such an instructional method that requires introductory statistics students to follow blindly fixed procedural steps, with no regard for the rationale underlying these steps, does very little to contribute to the students' statistics educational experience.

It is also important to remember that the ultimate purpose of conducting statistical analysis is 'interpretation' – correctly interpreting the statistical results in relation to the researcher's hypotheses. Correct interpretation of the test results is not dependent on whether the student/researcher employs the manual cookbook technique or some other more efficient techniques such as specialized statistical software programs. Thus, in a practical sense, it does not matter how the results are derived as long as the results are correct and are then interpreted correctly.

To this end, this book proposes a different approach to learning statistical analyses for introductory social sciences students in which beginner students bypass the conventional manual cookbook method of learning and focus instead on how to conduct statistical analyses via the popular SPSS statistical program. Learning how to use this very useful and sophisticated statistical software package at the introductory class level has a number of distinct advantages. First, as mentioned earlier, the conventional cookbook method of instruction simply requires the first-year student to follow blindly a set of procedural steps; while such a technique may yield the correct results, the step-by-step calculations are simply not meaningful to a lot of students and contribute very little to their understanding of the rationale of how the correct results were obtained. Since the main focus of the analysis is on the interpretation of the obtained results (and not on the technique used to derive the results), it does not matter how the results were obtained as long as the obtained results are correct. In other words, it would be much more efficient to bypass the conventional cookbook method and to learn how to conduct statistical analysis via software packages such as SPSS.

A second advantage of learning the SPSS software package at the introductory class level is that most social sciences students will employ this program in their later years of study. This is because SPSS is one of the most popular of the many statistical packages currently available. Learning how to use this program at the very start of their academic journey not only familiarizes students with the utility of this program but also provides them with the experience to employ the program to conduct complex analyses in their later years of study.

2.2 Introduction to SPSS

When SPSS, Inc., was conceived in 1968, it stood for *Statistical Package for the Social Sciences.* Since the company's purchase by IBM in 2009, IBM decided to simply use the name SPSS (*Statistical Product and Service Solutions*) to describe its core product of predictive analytics. IBM describes predictive analytics as tools that *help connect data to effective action by drawing reliable conclusions about current conditions and future events.*

SPSS is an integrated system of computer programs designed for the analysis of social sciences data. It is one of the most popular of the many statistical packages currently available for statistical analysis. Its popularity stems from the fact that the program:

- Allows for a great deal of flexibility in the format of data
- Provides the user with a comprehensive set of procedures for data transformation and file manipulation
- Offers the researcher a large number of statistical analyses commonly used in the social sciences

For both the beginner and the advanced researchers, SPSS is an indispensable and powerful program that is relatively easy to use once the researcher has been taught the rudiments. The Windows version of SPSS has introduced a *point-and-click* interface that allows the researcher to merely point and click through a series of windows and dialogue boxes to specify the analysis required and the variables involved. This method eliminates the need to learn the very powerful syntax or command language used to execute SPSS analyses (in the older MS-DOS versions), and has proven to be highly popular for those researchers with little or no interest in learning the *syntax* method. Nevertheless, SPSS for Windows has retained the syntax method, which permits the researcher to execute SPSS analyses by typing commands (syntax files).

A question that is often asked by the beginner researcher is which method of conducting SPSS is better. Both the Windows method and the syntax method have their advantages and disadvantages, and these will be discussed in Section 2.3.

This chapter has been written with the beginner student and researcher in mind, and provides an overview of the two most basic functions of SPSS: (1) how to set up data files in SPSS for Windows and (2) conducting SPSS analysis via the Windows method and the syntax method.

2.2.1 Setting Up a Data File

Suppose a researcher has conducted a survey to investigate the extent of support by Thai people for four types of euthanasia (mercy killing) – active euthanasia, passive euthanasia, voluntary euthanasia, and non-voluntary euthanasia. Responses to these four death issues were obtained from the questionnaire presented in Table 2.1.

2.2.2 Preparing a Codebook

Prior to data entry, it will be useful to prepare a codebook that contains the names of the variables in the questionnaire, their corresponding SPSS variable names, and their coding instructions. An important purpose of such a codebook is to allow the researcher to keep track of all the variables in the survey questionnaire and the way they are defined in the SPSS data file. Table 2.2 presents the codebook for the questionnaire presented in Table 2.1.

2.2.3 Data Set

Table 2.3 presents the responses (raw data) obtained from a sample of 10 respondents to the euthanasia questionnaire.

2.2.4 Creating SPSS Data File

The following steps demonstrate how the data presented in Table 2.3 are entered into a SPSS data file.

1. When the SPSS program (Version 23) is launched, the following window will open.

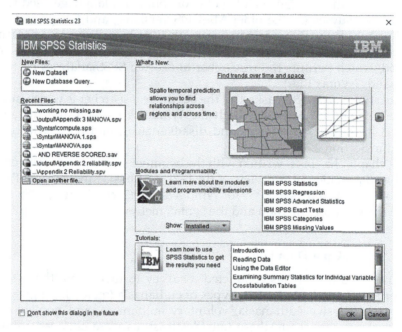

Since the purpose of the present exercise is to create a new data file, close this window by clicking [Cancel]. The following **Untitled1 [DataSet0] – IBM SPSS Statistics Data Editor** screen will then be displayed.

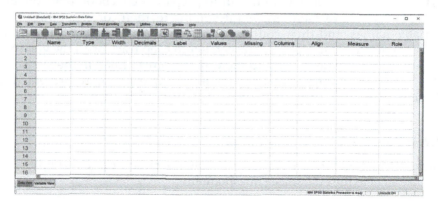

TABLE 2.1

Euthanasia Survey Questionnaire

---

a) Gender   1. _____ Male   2. _____ Female

b) Age _____ (in years)

c) The following 4 statements relate to the level of support for the 4 death issues of active euthanasia, passive euthanasia, voluntary euthanasia, and non-voluntary euthanasia. Please consider these 4 statements carefully and then decide your level of support for each of these 4 death issues. Please indicate your level of support by circling the number on each 6-point scale.

(i) Active euthanasia

1 _____ 2 _____ 3 _____ 4 _____ 5 _____ 6

| Strongly not support | Moderately not support | Barely not support | Barely support | Moderately support | Strongly support |

(ii) Passive euthanasia

1 _____ 2 _____ 3 _____ 4 _____ 5 _____ 6

| Strongly not support | Moderately not support | Barely not support | Barely support | Moderately support | Strongly support |

(iii) Voluntary euthanasia

1 _____ 2 _____ 3 _____ 4 _____ 5 _____ 6

| Strongly not support | Moderately not support | Barely not support | Barely support | Moderately support | Strongly support |

(iv) Non-voluntary euthanasia

1 _____ 2 _____ 3 _____ 4 _____ 5 _____ 6

| Strongly not support | Moderately not support | Barely not support | Barely support | Moderately support | Strongly support |

---

2. Prior to data entry, the variables in the data set must be named and defined. In the **Untitled1 [DataSet0] – IBM SPSS Statistics Data Editor** screen, the names of the variables are listed down the side (under the **Name** column), with their characteristics listed along the top (**Type, Width, Decimals, Label, Values, Missing, Columns, Align, Measure, Role**). The codebook presented in Table 2.2 will serve as a guide in naming and defining the variables. For example, the first variable is named **GENDER** and is coded 1 = male and 2 = female. Thus, in the first cell under **Name** in the **Data Editor** screen, type in the name **GENDER**. To assign the coded values (1 = male, 2 = female) to this variable, click the corresponding cell under **Values** in the **Data Editor** screen. Click the shaded area to open the following **Value Labels** window.

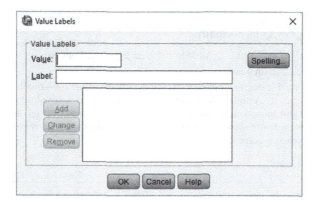

TABLE 2.2

Codebook

	SPSS	
Variable	**Variable Name**	**Code**
Gender	**Gender**	1 = male
		2 = female
Age	**Age**	Age in years
Active euthanasia	**Active**	1 = strongly not support
		2 = moderately not support
		3 = barely not support
		4 = barely support
		5 = moderately support
		6 = strongly support
Passive euthanasia	**Passive**	1 = strongly not support
		2 = moderately not support
		3 = barely not support
		4 = barely support
		5 = moderately support
		6 = strongly support
Voluntary euthanasia	**Voluntary**	1 = strongly not support
		2 = moderately not support
		3 = barely not support
		4 = barely support
		5 = moderately support
		6 = strongly support
Non-voluntary euthanasia	**Non-Voluntary**	1 = strongly not support
		2 = moderately not support
		3 = barely not support
		4 = barely support
		5 = moderately support
		6 = strongly support

TABLE 2.3

Raw Data

Gender	Age	Active Euthanasia	Passive Euthanasia	Voluntary Euthanasia	Non-Voluntary Euthanasia
1	24	2	3	6	1
1	18	2	4	5	1
2	33	1	3	6	2
1	29	2	4	5	1
2	26	1	4	5	2
2	19	3	4	4	2
1	36	2	3	6	1
2	34	2	3	6	1
1	20	1	4	5	2
2	21	2	5	6	1

3. In order to define the code for male respondents, type **1** in the **Value:** cell, and in the **Label:** cell, type **Male**. Next, click [Add] to complete the coding for the male respondents. For female respondents, type **2** in the **Value:** cell, and in the **Label:** cell, type **Female**. Next, click [Add] to complete the coding for the female respondents. The completed **Value Labels** window is presented below.

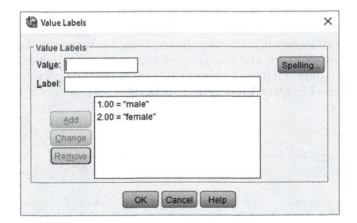

Next, click [OK] to complete the coding for the GENDER variable and to return to the **Untitled1 [DataSet0] – IBM SPSS Statistics Data Editor** screen below.

4. Repeat the above coding procedure for the rest of the variables in the codebook. Please note that the AGE variable is a *continuous* variable and therefore has no coded values.

5. If the researcher wishes to attach a label to a variable name (to provide a longer description for that variable), this can be done by typing a label in the corresponding cell in the **Label** column. For example, the researcher may wish to attach the label '**support for active euthanasia**' to the variable **ACTIVE**. This label will be printed in the analysis output generated by SPSS. The following **Untitled1 [DataSet0] – IBM SPSS Statistics Data Editor** screen displays the names of all the variables listed in the code book, and where relevant, their **Labels** and **Values** codes.

	Name	Type	Width	Decimals	Label	Values	Missing	Columns	Align	Measure	Role
1	gender	Numeric	8	2		{1.00, male}...	None	8	Right	Nominal	Input
2	age	Numeric	8	2		None	None	8	Right	Scale	Input
3	active	Numeric	8	2	support for active euthanasia	{1.00, stron...	None	8	Right	Scale	Input
4	passive	Numeric	8	2	support for passive euthanasia	{1.00, stron...	None	8	Right	Scale	Input
5	voluntary	Numeric	8	2	support for voluntary euthanasia	{1.00, stron...	None	8	Right	Scale	Input
6	non_voluntary	Numeric	8	2	support for non-voluntary euthanasia	{1.00, stron...	None	8	Right	Scale	Input

2.2.5 Data Entry

Data can only be entered via the **Data View** screen. Switch the present **Variable View** to **Data View** by clicking the **Data View** tab `Data View` `Variable View` at the bottom left-hand corner of the screen. In the **Data View** screen, the rows represent the respondents, and the columns represent the variables. Beginning with the first data cell (row 1, column 1), type in the data presented in Table 2.3. The following **Data View** screen shows the data obtained from the 10 respondents.

Chapter 2 example for data entry.sav [DataSet2] - IBM SPSS Statistics Data Editor

File Edit View Data Transform Analyze Direct Marketing Graphs Utilities Add-ons Window Help

11 : active   Visible: 6 of 6 Variables

	gender	age	active	passive	voluntary	non_voluntary
1	1.00	24.00	2.00	3.00	6.00	1.00
2	1.00	18.00	2.00	4.00	5.00	1.00
3	2.00	33.00	1.00	3.00	6.00	2.00
4	1.00	29.00	2.00	4.00	5.00	1.00
5	2.00	26.00	1.00	4.00	5.00	2.00
6	2.00	19.00	3.00	4.00	4.00	2.00
7	1.00	36.00	2.00	3.00	6.00	1.00
8	2.00	34.00	2.00	3.00	6.00	1.00
9	1.00	20.00	1.00	4.00	5.00	2.00
10	2.00	21.00	2.00	5.00	6.00	1.00

2.2.6 Saving and Editing Data File

Once data entry is completed, the data file can be saved. From the menu bar, click **File**, then **Save As**. Once it has been decided where the data file is to be saved to, type a name for the file. As this is a data file, SPSS will automatically append the suffix.**SAV** to the data file name (e.g., **TRIAL.SAV**).

To edit an existing file, click **File**, then **Open**, and then **Data** from the menu bar. Scroll through the names of the data files and double-click on the data file to open it.

2.3 SPSS Analysis: Windows Method versus Syntax Method

Once the SPSS data file has been created, the researcher can conduct the chosen analysis either through the Windows method (point-and-click) or the syntax method. The primary advantage of using the Windows method is clearly its ease of use. With this method, the researcher accesses the pull-down menu by clicking **Analyze** in either the **Data View** or **Variable View** mode, and then point-and-clicks through a series of windows and dialogue boxes to specify the kind of analysis required and the variables involved. There is no need to type in any syntax or commands to execute the analysis. Although this procedure seems ideal at first, paradoxically it is not always the method of choice for the more advanced and sophisticated users of the program. Rather, there is clearly a preference for the syntax method among these users. This preference stems from several good reasons from learning to use the syntax method.

First, when conducting complex analysis, the ability to write and edit syntax commands is advantageous. For example, if a researcher mis-specifies a syntax command for a complex analysis and wants to go back and rerun it with minor changes, or if the researcher wishes to repeat an analysis multiple times with minor variations, it is often more efficient to write and edit the syntax command directly than to repeat the Windows pull-down menu sequences. Second, from my teaching experience with SPSS, I believe that students have a better "feel" for statistics if they have to write syntax commands to generate the specific statistics they need, rather than merely relying on pull-down menus. In other words, it provides a better learning experience. Finally, and perhaps most important, several SPSS procedures are available only via the syntax method.

2.4 SPSS Analysis: Windows Method

Once the data have been entered, the researcher can begin the data analysis. Suppose the researcher is interested in obtaining general **descriptive statistics** for all of the variables entered in the data set **TRIAL.SAV**.

1. From the menu bar, click **Analyze**, then **Descriptive Statistics**, and then **Frequencies.** The following **Frequencies** window will open.

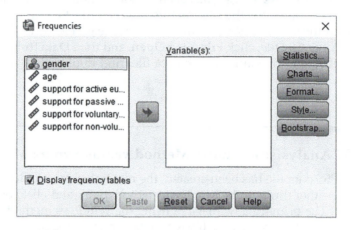

2. In the left-hand field containing the study's six variables, click (highlight) these variables, and then click to transfer the selected variables to the **Variable(s):** field.

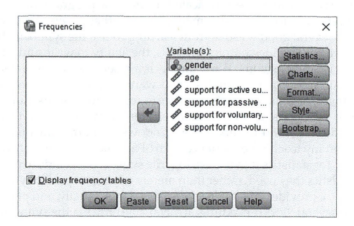

3. Click Statistics... to open the **Frequencies: Statistics** window below. Suppose the researcher is only interested in obtaining statistics for the **Mean, Median, Mode,** and **Standard Deviation** for the six variables. In the **Frequencies: Statistics** window, check the fields related to these statistics. Next click Continue.

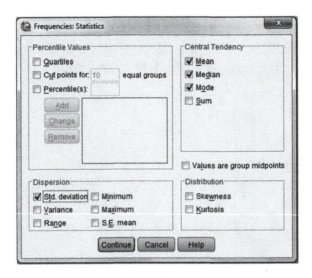

4. When the **Frequencies** window opens, run the analysis by clicking
   OK . See Table 2.4 for the results.

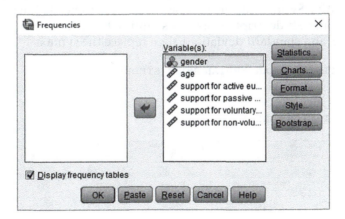

2.5 SPSS Analysis: Syntax Method

```
FREQUENCIES VARIABLES=ALL
/STATISTICS=MEAN MEDIAN MODE STDDEV.
```

1. From the menu bar, click **File**, then **New**, and then **Syntax.** The following **IBM SPSS Statistics Syntax Editor** window will open.

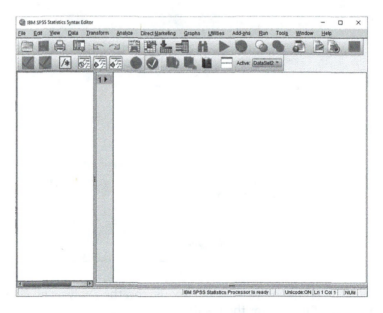

2. Type the **Frequencies** analysis syntax command in the **IBM SPSS Statistics Syntax Editor** window. If the researcher is interested in obtaining all descriptive statistics (and not just the mean, median, mode, and standard deviation), then replace the syntax:

```
/STATISTCS=MEAN MEDIAN MODE STDDEV.
```

with

```
/STATISTICS=ALL.
```

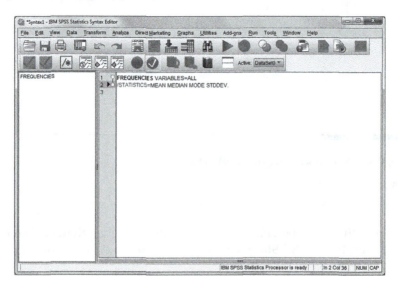

3. To run the Frequencies analysis, click ▶ or click <kbd>Run</kbd> and then **All**.

(NOTE: **Table E presents a glossary of the SPSS syntax files employed for all the examples in this book**).

2.5.1 SPSS Output

TABLE 2.4

Frequencies Output

		Gender	Age	Support for Active Euthanasia	Support for Passive Euthanasia	Support for Voluntary Euthanasia	Support for Non-Voluntary Euthanasia
				Frequencies			
				Statistics			
N	Valid	10	10	10	10	10	10
	Missing	0	0	0	0	0	0
Mean		1.5000	26.0000	1.8000	3.7000	5.4000	1.4000
Median		1.5000	25.0000	2.0000	4.0000	5.5000	1.0000
Mode		1.00[a]	18.00[a]	2.00	4.00	6.00	1.00
Std. Deviation		0.52705	6.66667	0.63246	0.67495	0.69921	0.51640

[a] Multiple modes exist. The smallest value is shown.

Frequency Table

Gender

		Frequency	Percent	Valid Percent	Cumulative Percent
Valid	Male	5	50.0	50.0	50.0
	Female	5	50.0	50.0	100.0
	Total	10	100.0	100.0	

Age

		Frequency	Percent	Valid Percent	Cumulative Percent
Valid	18.00	1	10.0	10.0	10.0
	19.00	1	10.0	10.0	20.0
	20.00	1	10.0	10.0	30.0
	21.00	1	10.0	10.0	40.0
	24.00	1	10.0	10.0	50.0
	26.00	1	10.0	10.0	60.0
	29.00	1	10.0	10.0	70.0
	33.00	1	10.0	10.0	80.0
	34.00	1	10.0	10.0	90.0
	36.00	1	10.0	10.0	100.0
	Total	10	100.0	100.0	

(Continued)

TABLE 2.4 (*Continued*)

Frequencies Output

	Support for Active Euthanasia			
		Frequency	Percent	Valid Percent	Cumulative Percent
Valid	Strongly not support	3	30.0	30.0	30.0
	Moderately not support	6	60.0	60.0	90.0
	Barely not support	1	10.0	10.0	100.0
	Total	10	100.0	100.0	

	Support for Passive Euthanasia			
		Frequency	Percent	Valid Percent	Cumulative Percent
Valid	Barely not support	4	40.0	40.0	40.0
	Barely support	5	50.0	50.0	90.0
	Moderately support	1	10.0	10.0	100.0
	Total	10	100.0	100.0	

	Support for Voluntary Euthanasia			
		Frequency	Percent	Valid Percent	Cumulative Percent
Valid	Barely support	1	10.0	10.0	10.0
	Moderately support	4	40.0	40.0	50.0
	Strongly support	5	50.0	50.0	100.0
	Total	10	100.0	100.0	

	Support for Non-Voluntary Euthanasia			
		Frequency	Percent	Valid Percent	Cumulative Percent
Valid	Strongly not support	6	60.0	60.0	60.0
	Moderately not support	4	40.0	40.0	100.0
	Total	10	100.0	100.0	

2.5.2 Results and Interpretation

The **Statistics** Table presents the requested Mean, Median, Mode, and Standard Deviation statistics for the six variables. The **Gender** variable is a nominal (categorical) variable and as such, its mean, median, and standard deviation statistics are not meaningful. The remaining five variables of **AGE, ACTIVE, PASSIVE, VOLUNTARY,** and **NON-VOLUNTARY** are measured at least at the ordinal level (i.e., they are continuous variables), and as such their mean, median, and standard deviation statistics can be interpreted.

The results presented in the **Statistics** Table show that the 10 respondents in the survey have a mean age of 26 years and a median age of 25 years. Since there is no one age that occurs more frequently than others, SPSS presents the lowest age value of 18 as the mode. The standard deviations (i.e., the average spread of the scores from the mean) for the four euthanasia variables are as follows: ACTIVE: $SD = 0.63$; PASSIVE: $SD = 0.67$; VOLUNTARY: $SD = 0.70$; NON-VOLUNTARY: $SD = 0.52$. For the AGE variable, the standard deviation shows that its average deviation from the mean is 6.67 years.

For the four variables of 'support for' active, passive, voluntary, and non-voluntary euthanasia, the results show that support for voluntary euthanasia is highest (Mean = 5.40; Median = 5.50), followed by passive euthanasia (Mean = 3.70; Median = 4.00), active euthanasia (Mean = 1.80; Median = 2.00), and non-voluntary euthanasia (Mean = 1.40; Median = 1.00).

The **Frequency Table** presents the breakdown of the frequency distributions for the six variables (**GENDER, AGE, ACTIVE, PASSIVE, VOLUNTARY,** and **NON-VOLUNTARY**). For each variable, the Frequency Table presents (1) the **Frequency** of occurrence for each value within that variable, (2) the frequency for each value expressed as a **Percentage** of the total sample, (3) the **Valid Percentage** for each value, controlling for missing cases, and (4) the **Cumulative Percentage** for each succeeding value within that variable. For example, the Frequency Table for the variable of **Gender** shows that there are 5 males and 5 females in the sample, and that these two groupings represent 50% each of the total sample. Since there are no missing cases, the valid percentage values are identical to the percentage values. *If there are missing cases, then the valid percentage values should be interpreted.* The cumulative percentage presents the percentage of scores falling at or below each score. Thus, for the sample of 10 respondents, the 5 males in the sample represent 50% of the sample, and the additional 5 females represent a cumulative percentage of 100%.

The frequency Tables for the variables of **AGE, ACTIVE, PASSIVE, VOLUNTARY,** and **NON-VOLUNTARY** are interpreted in exactly the same way. Taking support for ACTIVE euthanasia as an example, it can be seen that 3 respondents (30%; 30 cumulative percent) responded with 'strongly not support', 6 respondents (60%; 90 cumulative percent) responded with 'moderately not support', and 1 respondent (10%; 100 cumulative percent) responded with 'barely not support'.

Section I

Descriptive Statistics

3

Basic Mathematical Concepts and Measurement

3.1 Basic Mathematical Concepts

Let's start off by being honest! Many students in the social sciences are not noted for their mathematical prowess. Indeed, many students approach statistics with a great deal of anxiety. Yet, I believe that a lot of this anxiety is unnecessary, and the belief that one has to master many mathematical formulas can only serve to exacerbate this anxiety. In reality though, in learning statistics one does not have to be a mathematical genius to master basic statistical principles. In other words, one does not have to be an expert in calculus or differential equations to do statistics. Yet, we cannot totally do away with mathematics as statistics requires a good deal of arithmetic computation. To be successful in a statistics course, the student must be familiar with elementary algebra and basic mathematical operations (addition, subtraction, multiplication, division, square, and square root), which most students learned during their high school years.

3.1.1 Mathematical Notations

A basic fact in statistics is that the procedure employs numerous symbols, and it is to the student's advantage to learn/memorize these symbols. Mastering these symbols makes understanding the material that much easier. While there are many statistical symbols that students will encounter in their statistics journey, there are four that appear with great regularity. These include X, Y, N, and Σ.

- The symbols X and Y are used to represent the variables measured. Thus, if IQ and age were two variables measured in a study, X might be used to represent IQ and Y to represent age. Table 3.1 shows the distribution of these two variables for 5 subjects.

TABLE 3.1

Distribution of IQ (*X*) and Age (*Y*) Scores

Subject Number (N)	Variable Symbol IQ (X)	Variable Scores (X)	Variable Symbol Age (Y)	Variable Scores (Y)
1	X_1	118	Y_1	22
2	X_2	105	Y_2	19
3	X_3	122	Y_3	16
4	X_4	125	Y_4	20
5	X_5	112	Y_5	21

For this example, each of the 5 *X* scores and each of the 5 *Y* scores represent a specific value of IQ (*X*) and age (*Y*). These scores are differentiated by their subscripts such that each score corresponds to a particular subject that had the specific value. Thus, the score symbol X_1 denotes the IQ score (118) of the first subject; X_2 denotes the IQ score (105) of the second subject, and so on. Similarly, the score symbol Y_1 denotes the age score (22) of the first subject; X_2 denotes the age score (19) of the second subject, and so on.

- Another frequently used mathematical notation is the symbol *N*, which represents the number of subjects in the data set. Thus, for the example above, there are 5 subjects,

$$N = 5$$

- A symbol that is used most frequently to denote the summation of all or part of the scores in the data set is Σ(sigma). The full algebraic summation phrase is

$$\sum_{i=1}^{N} X_i$$

The notations above (*N*) and below (*i* = 1) the summation sign indicate which scores to include in the summation. The notation below the summation sign tells us the first score in the summation, and the notation above the summation sign indicates the last score. Thus, this algebraic summation phrase tells us to sum the *X* scores beginning with the first score (X_1) and ending with the N_{th} (last) score (X_5). Applying this to the IQ data,

$$\sum_{i=1}^{N} Xi = X_1 + X_2 + X_3 + X_4 + X_5$$
$$= 118 + 105 + 122 + 125 + 112$$
$$= 582$$

When the summation is across all the scores in the distribution, then the full algebraic summation phrase can be summarized as

$$\sum X = 582$$

Note that it is not necessary to sum across all the scores in the distribution, that is, from 1 to N. For example, we may want to sum only selected quantities, say the second, third, and fourth scores. Thus, the algebraic summation phrase would be

$$\sum_{i=2}^{4} X_i = X_2 + X_3 + X_4$$
$$= 105 + 122 + 125$$
$$= 352$$

3.2 Measurement Scales (Levels of Measurement)

In conducting a statistical analysis, the researcher must consider the measurements of the variables to be analyzed. Different variables may be measured differently, that is, on different scales. For example, to measure reaction time in milliseconds, we might use a stop watch. But stop watches are of no use when they come to measuring a person's IQ, age, or height. Determining how variables are measured is important because (1) it helps the researcher interpret the data obtained from the measurement of that variable and (2) the levels at which the variables are measured determine which statistical test is used to analyze the data. Most typically, variables in the behavioral sciences are measured on one of four scales: *nominal, ordinal, interval,* or *ratio* measurements. These four types of scales differ in the number of the following attributes they possess: *magnitude, an equal interval between adjacent units,* and *an absolute zero point.*

3.2.1 Nominal Scales

Nominal scales embody the lowest level of measurement because the measurement involves simply categorizing the variable to be measured into one of a number of discrete categories. Gender, hair color, religion, and ethnicity are examples of variables measured on a nominal scale. The measurement of these variables does not imply any ordering or magnitude among the responses. For instance, in measuring ethnicity, people may be categorized as American, Chinese, Australian, African, or Indian. Once people have been categorized into these categories, all people in a particular category (e.g., those categorized as Americans) are assumed to be equated on the measurement obtained, even though they may not possess the same number of the characteristics. Numbers can be assigned to describe the categories (e.g., 1 = American, 2 = Chinese, 3 = Australian, 4 = African, and 5 = Indian), but the numbers are only used to name/label the categories. They have no magnitude in terms of quantitative value.

3.2.2 Ordinal Scales

This level of measurement involves ordering or ranking the variable to be measured. For example, a researcher may be interested in finding out how satisfied customers are with their new refrigerator. They are asked to specify their level of satisfaction on a 4-point scale ranging from '1 = very dissatisfied', '2 = dissatisfied', '3 = satisfied', or '4 = very satisfied'. This range indicates that the reported level of satisfaction is ordered, ranging from least satisfied (a rank of 1) to most satisfied (a rank of 4). These numbers allow the researcher to quantify the magnitude of the measured variable, by adding the arithmetic relationships "greater than" and "less than" to the measurement process. While ordinal scales allow one to differentiate between rankings among the variable being measured, they do not permit determination of how much of a real difference exists in the measured variable between ranks. Thus, a customer who ranked his level of satisfaction as 4 (very satisfied) is considered to be more satisfied with the new refrigerator than a customer who ranked his level of satisfaction as 3 (satisfied). But these rankings provide no information on how much more satisfied the customer who ranked 4 is over the customer who ranked 3. In other words, the intervals between the ranks are not meaningful.

Ordinal scales also fail to capture another piece of important information, that is, the difference between two rankings on an ordinal scale cannot be assumed to be equal to the difference between two other rankings. In our satisfaction scale, for example, the difference between the responses 'very dissatisfied' and 'dissatisfied' is probably not equivalent to the difference between 'dissatisfied' and 'satisfied'. That is, nothing in the ordinal scale allows us to determine whether the differences in the two rankings reflect the same difference in psychological satisfaction.

3.2.3 Interval Scales

This level of measurement involves being able to specify how far apart two stimuli are on a given dimension. That is, the intervals on the scale have the same interpretation throughout. As mentioned earlier, on an ordinal scale, the difference between 'level of satisfaction' ranked 4 (very satisfied) and satisfaction ranked 3 (satisfied) does not necessarily equal the distance between satisfaction ranked 1 (very dissatisfied) and 2 (dissatisfied). On an interval scale, however, differences of the same numerical size in scale values are equal. For example, for scores obtained from a statistics exam, a 40-point difference in exam scores has the same meaning anywhere along the scale. Thus, the difference in exam test scores between 80 and 40 is the same as the difference between 60 and 20. However, it would not be correct to say that a person with an exam score of 80 is twice as good in statistics as a person with a score of 40. The reason for this is because exam scores (and other similar interval scales) do not have a true zero that represents a complete absence of statistical skills. That is, a student who obtained an exam score of '0' does not mean that that student is totally devoid of statistical knowledge. Perhaps that particular statistics exam was simply too difficult for that student and that he/she would have done better on another easier exam.

3.2.4 Ratio Scales

This level of measurement replaces the arbitrary zero point of the interval scale with a true zero starting point that corresponds to the absence of the variable being measured. The ratio scale of measurement is the most informative scale in that it combines all the characteristics of the three earlier scales. It is a nominal scale in that it provides a name or category for each object (the numbers serve as labels). It is an ordinal scale in that the objects are ordered (in terms of the ranking of the numbers). It is an interval scale in that the same difference at two places on a scale has the same meaning. It is also a ratio scale in that the same ratio at two places on the scale carries the same meaning. Thus, with a ratio scale it is possible to state that a variable has twice, or half, or three times as much of the variable measured than another. Take weight as an example. Weight has a true zero point (a weight of zero means that the object is weightless) and the intervals between the units of measurement are equal. Thus, the difference between 10 and 15 g is equal to the difference between 45 and 50 g, and 80 g is twice as heavy as 40 g.

3.3 Types of Variables

3.3.1 IV and DV

When a researcher conducts an experiment, the variables that are tested can be classified as *IVs* and *DVs*. An IV is the variable that is manipulated or controlled in a scientific experiment to test its effects on the DV. A DV is the variable affected by the manipulation of the IV. That is, the DV is 'dependent' on the IV and as the experimenter manipulates/changes the IV, the effect on the DV is observed and recorded.

For example, a developmental psychologist may be interested in finding out whether gender has any effect on problem-solving skill. The psychologist manipulates 'gender' by setting up two groups – males and females. Thus, 'gender' is the IV. The problem-solving scores observed and recorded for the two groups of males and females would be the DV.

3.3.2 Continuous and Discrete Variables

While variables can be distinguished between IV and DV, they can also be differentiated as *continuous* or *discrete*. A continuous variable is one that potentially has an infinite number of possible values between adjacent units on a scale. Take weight as an example. Between 1 and 2 g, there can be an infinite number of values (e.g., 1.1, 1.2, 1.3, 1.689 g, etc.)

This is not the case with discrete variables. A discrete variable is a variable that can only take on a certain number of values. That is, there are no possible values between adjacent units on the scale. Take the number of cars on a road as an example. There may be 1 car, 2 cars, 3 cars, 4 cars, and so forth. But there can be no possible values between 1 car and 2 cars (e.g., 1.26 cars!) or between 2 cars and 3 cars (e.g., 2.56 cars!). The value of a discrete variable changes in fixed amounts with no intermediate values possible.

3.3.3 Real Limits of Continuous Variables

Recall that a continuous variable has an infinite number of possible values between adjacent units on a scale. Therefore, when we measure a continuous variable like time, it is necessary to understand that the measure is only an approximation. This is because of the possibility that a more sensitive measuring instrument can increase the accuracy of our measurements a little more. In general, *the real limits of a number equals the number plus and minus ($\pm$) 1/2 the unit of measurement.*

Examples,

Number	Unit of Measurement	1/2 Unit of Measurement	Real Limits	
			Lower	Upper
13.3	0.1	0.05	13.25	13.35
13.33	0.01	0.005	13.325	13.335
13.333	0.001	0.0005	13.3325	13.3335
13	1	0.5	12.5	13.5

Take weight as an example. Let us say you step on a bathroom scale (calibrated in 1 kilogram units) and the scale-pointer points to your weight as 56.6 kg. Since weight is a continuous variable, this value is only an approximate. However, it is possible to specify the limits/interval within which the true value falls. Since the true limits of a value of a continuous variable is

TABLE 3.2

Rounding Numbers Up or Down to
Two Decimal Places

Before Rounding	After Rounding
3.546	3.55
3.873	3.87
5.659	5.66
7.863	7.86

defined as being *equal to that number plus or minus* (±) *one half of the unit of measurement*, your true weight must be within the range 56.1 and 57.1 kg (i.e., 56.6 ± 0.5). These numbers are called the real limits of that measure.

3.3.4 Rounding

After conducting an analysis, the researcher has to report the results. When reporting numerical results, two questions are asked: (1) How many decimal places does the final answer carry? (2) What value should the last digit have? The rules to be followed in answering these two questions are both simple and straightforward.

First, in answering the question of how many decimal places the final answer should carry, it is up to the researcher to decide whether to report the final answer to two, three, or four decimal places.

Second, in answering the question of what value should the last digit have, follow these steps:

- Decide which is the last digit to *keep*
- Leave it the same if the *next digit* is less than 5 (this is called *rounding down*)
- Increase it by 1 if the next digit is 5 or more (this is called *rounding up*)

Table 3.2 shows how numbers are rounded up or down to two decimal places.

4

Frequency Distributions

4.1 Ungrouped Frequency Distributions

Frequency distributions are visual displays that *organize* and present *frequency counts* so that the information can be interpreted more easily. Frequency distributions show the number of times a given quantity (or group of quantities) occurs in a set of data. For example, the frequency distribution of age in a population would show how many individuals are at a certain age level (say, 26 years). In considering the frequency distribution of age, the *frequency* (*f*) of a certain age is the number of times that age occurs in the data. The *distribution* (*d*) of age is the pattern of frequencies of the observation.

The primary purpose of frequency distributions is to assist the researcher to make sense of the data collected. Let's suppose that you are one of 100 individuals who took an IQ test and, your score is 120. A score of 120 is simply a number and is virtually meaningless unless you have a standard against which to compare your score. For example, how high (or low) is your IQ score relative to the other individuals who took the test? How many people received an IQ score higher than yours? How many scores were lower? Table 4.1 presents the raw IQ scores from all 100 individuals.

Although all 100 IQ scores are shown, they are presented randomly, which makes it difficult if not impossible to make sense of these scores. A more efficient way to present this data set and one which conveys more meaning is to list the 100 IQ scores in terms of their frequency of occurrence, that is, the scores *frequency distribution*.

To conduct a frequency analysis on the 100 IQ scores, choose either the **SPSS Windows** method or the **SPSS Syntax File** method.

TABLE 4.1

IQ Scores ($N = 100$)

155	111	136	128	124	116	135	104	86	113
134	132	114	88	109	127	118	90	114	121
117	120	123	128	90	107	120	111	144	108
129	104	116	109	95	109	92	123	81	109
86	94	104	113	140	116	138	124	103	112
101	109	91	101	83	134	102	125	133	101
106	101	109	118	114	115	108	126	97	98
151	112	115	122	117	120	113	116	130	112
119	131	128	106	110	105	122	114	84	132
98	105	88	137	110	148	126	129	143	110

4.1.1 SPSS: Data Entry Format

The data set has been saved under the name: **EX1.SAV**

Variable	Column(s)	Code
IQ	1	IQ scores
GROUP	2	$1 = 151–157$
		$2 = 144–150$
		$3 = 137–143$
		$4 = 130–136$
		$5 = 123–129$
		$6 = 116–122$
		$7 = 109–115$
		$8 = 102–108$
		$9 = 95–101$
		$10 = 88–94$
		$11 = 81–87$

4.1.2 SPSS Windows Method

1. When the SPSS program (Version 23) is launched, click **Analyze** on the menu bar, then **Descriptive Statistics**, and then **Frequencies**. The following **Frequencies** window will open.

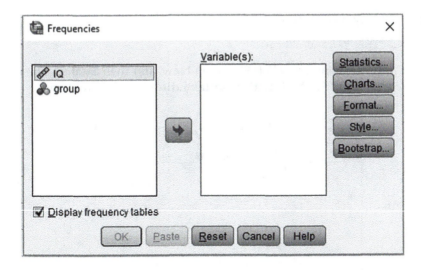

2. In the left-hand field containing the study's **IQ** variable, click (highlight) this variable, and then click 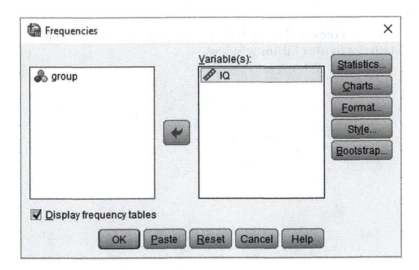 to transfer the selected IQ variable to the **Variable(s):** field.

3. As the purpose of this analysis is to obtain a frequencies distribution for the 100 IQ scores (i.e., you are not interested in obtaining statistics for the variable's mean, median, mode, and standard deviation), click ok to run the analysis. See Table 4.2 for the results.

4.1.3 SPSS Syntax Method

```
FREQUENCIES VARIABLES=IQ.
```

1. From the menu bar, click **File**, then **New**, and then **Syntax.** The following **IBM SPSS Statistics Syntax Editor** Window will open.

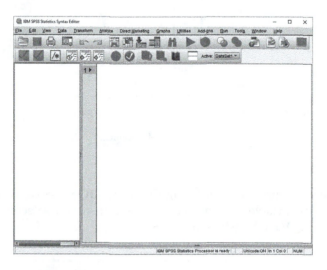

2. Type the **Frequencies** analysis syntax command in the **IBM SPSS Statistics Syntax Editor** window.

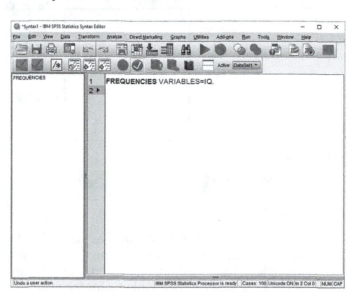

3. To run the Frequencies analysis, click ▶ or click **Run** and then **All.**

4.1.4 SPSS Output

TABLE 4.2

Frequencies Output

Frequencies				
Statistics				
IQ				
N	Valid	100		
	Missing	0		

IQ				
		Frequency	Percent	Valid Percent	Cumulative Percent
---	---	---	---	---	---
Valid	81.00	1	1.0	1.0	1.0
	83.00	1	1.0	1.0	2.0
	84.00	1	1.0	1.0	3.0
	86.00	2	2.0	2.0	5.0
	88.00	2	2.0	2.0	7.0
	90.00	2	2.0	2.0	9.0
	91.00	1	1.0	1.0	10.0
	92.00	1	1.0	1.0	11.0
	94.00	1	1.0	1.0	12.0
	95.00	1	1.0	1.0	13.0
	97.00	1	1.0	1.0	14.0
	98.00	2	2.0	2.0	16.0
	101.00	4	4.0	4.0	20.0
	102.00	1	1.0	1.0	21.0
	103.00	1	1.0	1.0	22.0
	104.00	3	3.0	3.0	25.0
	105.00	2	2.0	2.0	27.0
	106.00	2	2.0	2.0	29.0
	107.00	1	1.0	1.0	30.0
	108.00	2	2.0	2.0	32.0
	109.00	6	6.0	6.0	38.0
	110.00	3	3.0	3.0	41.0
	111.00	2	2.0	2.0	43.0
	112.00	3	3.0	3.0	46.0
	113.00	3	3.0	3.0	49.0
	114.00	4	4.0	4.0	53.0
	115.00	2	2.0	2.0	55.0
	116.00	4	4.0	4.0	59.0
	117.00	2	2.0	2.0	61.0
	118.00	2	2.0	2.0	63.0
	119.00	1	1.0	1.0	64.0
	120.00	3	3.0	3.0	67.0
	121.00	1	1.0	1.0	68.0
	122.00	2	2.0	2.0	70.0
	123.00	2	2.0	2.0	72.0

(Continued)

TABLE 4.2 (*Continued*)

Frequencies Output

		IQ		
	Frequency	Percent	Valid Percent	Cumulative Percent
124.00	2	2.0	2.0	74.0
125.00	1	1.0	1.0	75.0
126.00	2	2.0	2.0	77.0
127.00	1	1.0	1.0	78.0
128.00	3	3.0	3.0	81.0
129.00	2	2.0	2.0	83.0
130.00	1	1.0	1.0	84.0
131.00	1	1.0	1.0	85.0
132.00	2	2.0	2.0	87.0
133.00	1	1.0	1.0	88.0
134.00	2	2.0	2.0	90.0
135.00	1	1.0	1.0	91.0
136.00	1	1.0	1.0	92.0
137.00	1	1.0	1.0	93.0
138.00	1	1.0	1.0	94.0
143.00	1	1.0	1.0	95.0
144.00	1	1.0	1.0	96.0
148.00	1	1.0	1.0	97.0
149.00	1	1.0	1.0	98.0
151.00	1	1.0	1.0	99.0
155.00	1	1.0	1.0	100.0
Total	100	100.0	100.0	

4.1.5 Results and Interpretation

As can be seen from Table 4.2, the IQ scores have been rank-ordered from the lowest to the highest, with each score's *frequency count, percent, valid percent,* and *cumulative percent* presented.

- The *frequency count* is simply the number of times that score has occurred.
- The *percent* is that score's frequency of occurrence expressed as a percentage of the total scores' frequency, *inclusive of missing scores* (i.e., $N = 100$). Suppose that there are 4 IQ scores missing (i.e., there are only 96 valid cases). Thus, for the IQ score of 109 with a frequency count of 6, its percentage (6%) will be calculated on the basis of the total scores' frequency, inclusive of the 4 missing cases (i.e., $N = 100$).
- The *valid percent* is that score's frequency of occurrence expressed as a percentage of the total scores' frequency, *exclusive of missing scores.* Using the same example where there are 4 IQ scores missing (i.e., there are only 96 valid cases), the valid percentage for the IQ score of 109 with a frequency count of 6 will be calculated on the basis

of the total scores' frequency, exclusive of the 4 missing cases (i.e., $N = 96; 6.3\%$).

- The *cumulative percent* is used to determine the percentage of scores that lie above (or below) a particular value. For example, in Table 4.2, the IQ score of 91 has a cumulative percent of 10. This value is the sum of all the valid percent values $(1 + 1 + 1 + 2 + 2 + 2 + 1)$ up to and inclusive of the score of 91. Thus, this cumulative percent informs the researcher that the overall IQ scores from 81 to 91 inclusive represent 10% of the total 100 IQ scores.

Presented this way, the data set can be interpreted more meaningfully. For example, how many students scored higher/lower than your score of 120? By summing the appropriate frequency counts, it is easy to see that 33 students (cumulative percent = 33%) have IQ scores higher than yours and 64 students (cumulative percent = 64%) have IQ scores lower than yours. It is also easy to see that there are 3 students with IQ scores of 120. From this example, it can be seen that the primary purpose of a frequency distribution is to organize data in such a way as to facilitate understanding and interpretation.

4.2 Grouped Frequency Distributions

Table 4.2 contains the ungrouped frequency distribution of the 100 IQ scores. When there is a large number of scores and the scores are widely spread out (as is in this example), listing individual scores has resulted in many scores with a very low frequency count of 1, and a display that is difficult to visualize its measure of central tendency (a measure of central tendency is a summary measure that describes a whole set of data with a single value that represents the middle or centre of its distribution). Under these circumstances, it is useful to group the scores into a manageable number of intervals (around 10–20) by creating *class intervals* of equal widths and computing the frequency of scores that fall into each interval. Such a distribution is called a *frequency distribution of grouped scores*.

4.2.1 Grouping Scores into Class Intervals

Grouping scores into class intervals are akin to collapsing the entire distribution of individual scores into classes in which the interval width of the classes is defined. It should be noted that grouping inevitably results in the loss of information. The wider the class interval set, the more information is lost. For example, Table 4.3 presents the 100 IQ scores grouped together into 5 classes with an interval width of 20 units wide.

It can be seen that although an interval this wide has resulted in a display with no low frequency count of 1, a lot of information has been lost. First,

TABLE 4.3

IQ Scores from Table 4.2 Grouped into Class
Intervals of 20 Units Wide

Class Interval (width = 20)	Frequency
81–100	16
101–120	51
121–140	27
141–160	6

individual scores listed in Table 4.2 have lost their identity when they are
grouped into class intervals and some errors in statistical estimation based
on group scores are unavoidable. Second, there are 51 scores listed within the
interval 101–120. But how are these 51 scores distributed within this interval?
Do they all fall at 101? Or at 120? Or are they distributed evenly across the
interval? The answer is that we don't know as that information has been lost.
At this point, the question is, "How does the researcher decide what interval width to employ?" It is obvious that the interval selected must not be so
gross that the researcher loses the discrimination provided by the original
ungrouped frequency distribution. On the other hand, the class intervals
should not be so fine that the purposes served by grouping are defeated. To
get the best of both worlds, the researcher must choose an interval width that
is neither too wide nor too narrow. By convention, it is generally agreed that
most data in the behavioural sciences can be accommodated by 10–20 class
intervals. Within this range, the actual number of class intervals employed
depends on the number and range of the original scores, that is, the 100 IQ
scores.

4.2.2 Computing a Frequency Distribution of Grouped Scores

Let's suppose that we have decided to group the data (the 100 IQ scores) into
approximately 10 class intervals. The procedural steps to be employed are as
follows:

Step 1. *Find the range of the scores.* The range is simply the difference
between the highest and lowest score values contained in the original data. That is,

$$\text{Range} = \text{Highest score minus lowest score} = 155 - 81 = 74$$

Step 2. *Determine the interval width (i).* Since we want approximately 10
class intervals to represent the distribution of grouped scores, divide
the range with this figure.

$$i = \frac{\text{Range}}{\text{Number of class intervals}} = \frac{74}{10} = 7.4$$

As the resulting value is not a whole number, apply the rule of rounding. As the decimal point (0.4) is less than 0.5, we round down to the whole number of 7. Thus, in the present example, $i = 7$.

Step 3. *List the class intervals.* We begin with the lowest class interval. The minimum value (lower limit) of this lowest class interval is simply the lowest score (**81**) in the original data. Selecting the lowest score as the minimum value of the lowest class interval ensures that the interval contains the lowest score. The maximum value (upper limit) of this lowest class interval is calculated as: Minimum value $+ (i - 1) = 81 + (7 - 1) = 87$. Thus, the lowest class interval for this data set is **81–87**.

Once the lowest class interval for the data set has been found, we can list the rest of the intervals. The next higher class interval starts with the score value (lower limit) that follows the maximum score value of the previous lowest class interval, which is **88**. The maximum value (upper limit) of this higher class interval is therefore **94** $(88 + (7 - 1))$. Thus, the next higher class interval for this data set is **88–94**.

Follow these procedural steps to compute each successive higher class interval until all the scores in the data set are listed in their appropriate class intervals.

Step 4. *Assign raw scores to the class intervals.* Assign each of the 100 IQ scores (from Table 4.1) to the class interval within which it is included. Table 4.4 presents the *grouped frequency distribution* of the 100 IQ scores in Table 4.1.

From this grouped frequency distribution, we can obtain an overall picture of the distribution of the IQ scores. For example, we note that there is a clustering of scores between the class intervals 102–108 and 123–129 ($N = 63$). We also note that the number of scores at the extreme ends of the distribution tends to taper off.

TABLE 4.4

Grouped Frequency Distribution of the 100 IQ Scores in Table 4.1

Class Interval	Real Limits	Frequency (*f*)	Percent	Cumulative Percent
151–157	150.5–157.5	2	2	2
144–150	143.5–150.5	3	3	5
137–143	136.5–143.5	3	3	8
130–136	129.5–136.5	9	9	17
123–129	122.5–129.5	13	13	30
116–122	115.5–122.5	15	15	45
109–115	108.5–115.5	23	23	68
102–108	101.5–108.5	12	12	80
95–101	94.5–101.5	8	8	88
88–94	87.5–94.5	7	7	95
81–87	80.5–87.5	5	5	100
				$N = 100$

Table 4.4 also presents the *real limits* of each class interval. From our previous discussion on the real limits of continuous variables (see Section 3.3.3), an IQ score is a continuous variable and therefore your IQ score of 120 is only an approximate. However, it is possible to specify the limits/interval within which the true value falls. Since the true limits of a value of a continuous variable is defined as being *equal to that number plus or minus ($\pm$) one half of the unit of measurement*, your true IQ score must be within the class interval range of **115.5–122.5**. This range specifies the real limits of the class interval that contains your IQ score of 120. The value 115.5 ($116 - 0.5$) is called the lower real limit and 122.5 ($122 + 0.5$) is the upper real limit.

4.2.3 SPSS Method

The grouped frequency distribution of the 100 IQ scores presented in Table 4.4 can be generated via either the **SPSS Windows method** or the **SPSS Syntax method**. For both methods, the data file containing the 100 IQ scores must be set up in such a way that the class intervals are specified and coded accordingly. The procedural steps to do this are as follows:

Step 1. Access and open the data file **EX1.SAV.** Create a new variable called **GROUP,** which will be used to define the 11 class intervals presented in Table 4.4.

	Name	Type	Width	Decimals	Label	Values	Missing
1	IQ	Numeric	8	2		None	None
2	group	Numeric	8	2		{1.00, 151 -...	None
3							

Step 2. To assign the 11 coded values (representing the 11 class intervals) to this **GROUP** variable, click the corresponding cell under **Values** in the **Data Editor** screen. Click the shaded area to open the following **Value Labels** window.

Step 3. To define the codes for the 11 class intervals (see Table 4.4), type **1** in the **Value:** cell, and in the **Label:** cell, type the class interval **151–157**. Next, click [Add] to complete the coding for this class interval. For the next lower class interval, type **2** in the **Value:** cell, and in the **Label:** cell, type **144–150**. Next, click [Add] to complete the coding for this class interval. Do the same for the rest of the 9 class intervals. The completed **Value Labels** window is presented below.

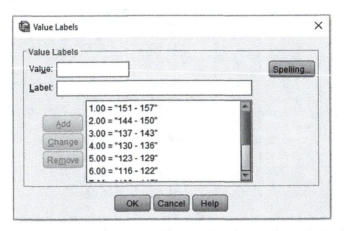

Next, click [OK] to complete the coding for the **GROUP** variable and to return to the **EX1.SAV [DataSet2] – IBM SPSS Statistics Data Editor** screen below.

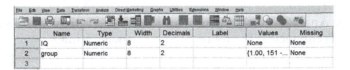

Step 4. Enter the **GROUP** values. Switch the present **Variable View** to **Data View** by clicking the **Data View** tab [Data View | Variable View] at the bottom left-hand corner of the screen. Note that in the **Data View** screen, the rows represent the respondents, and the columns represent the variables. For each respondent's IQ score, assign a **GROUP** code value to denote which class interval that IQ score is included. For example, the **Data View** screen shows that the first respondent has an IQ score of 155. Therefore, the **GROUP** code value for this score is recorded as **1** to denote that this IQ score falls within the class interval 151–157 (see Section 4.1.1). The second respondent has an IQ score of 134. Therefore, the **GROUP** code value for this score is recorded as **4** to denote that this IQ score falls within the class interval 130–136. The following

Data View screen shows the IQ scores and their corresponding **GROUP** code values (representing the class intervals) for the first 10 respondents.

File	Edit	View	Data	Transform	Analyze

	IQ	group
1	155.00	1.00
2	134.00	4.00
3	117.00	6.00
4	129.00	5.00
5	86.00	11.00
6	101.00	9.00
7	106.00	8.00
8	151.00	1.00
9	119.00	6.00
10	98.00	9.00

Once the data file has been set up correctly we can proceed with the analysis.

4.2.4 SPSS Windows Method

1. Click **Analyze** on the menu bar, then **Descriptive Statistics**, and then **Frequencies**. The following **Frequencies** window will open.

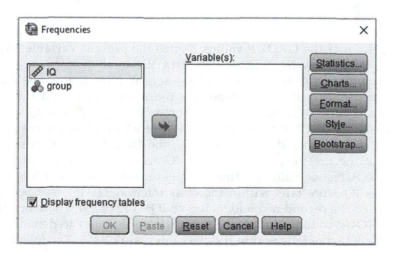

2. In the left-hand field containing the study's **GROUP** variable, click (highlight) this variable, and then click 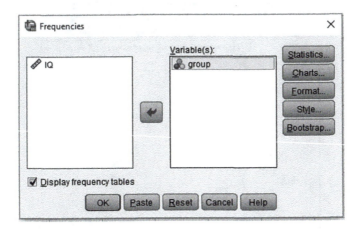 to transfer the selected **GROUP** variable to the **Variable(s):** field.

3. Click 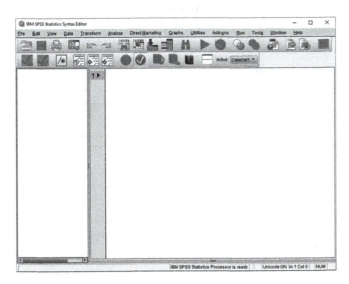 to run the analysis. See Table 4.5 for the results.

4.2.5 SPSS Syntax Method

```
FREQUENCIES VARIABLES=GROUP.
```

1. From the menu bar, click **File**, then **New**, and then **Syntax.** The following **IBM SPSS Statistics Syntax Editor** window will open.

2. Type the **Frequencies** analysis syntax command in the **IBM SPSS Statistics Syntax Editor** window.

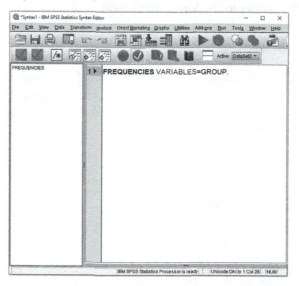

3. To run the Frequencies analysis, click ▶ or click Run and then **All**.

4.2.6 SPSS Output

TABLE 4.5

Grouped Frequency Distribution of the 100 IQ Scores Generated by SPSS

Frequencies				
Statistics				
Group				
N	Valid	100		
	Missing	0		

		Group			
		Frequency	**Percent**	**Valid Percent**	**Cumulative Percent**
Valid	151–157	2	2.0	2.0	2.0
	144–150	3	3.0	3.0	5.0
	137–143	3	3.0	3.0	8.0
	130–136	9	9.0	9.0	17.0
	123–129	13	13.0	13.0	30.0
	116–122	15	15.0	15.0	45.0
	109–115	23	23.0	23.0	68.0

(Continued)

TABLE 4.5 (*Continued*)

Grouped Frequency Distribution of the 100 IQ Scores Generated by SPSS

	Frequency	Percent	Valid Percent	Cumulative Percent
			Group	
102–108	12	12.0	12.0	80.0
95–101	8	8.0	8.0	88.0
88–94	7	7.0	7.0	95.0
81–87	5	5.0	5.0	100.0
Total	100	100.0	100.0	

It can be seen that the grouped frequency distribution of the 100 IQ scores generated by SPSS (Table 4.5) is identical to the grouped frequency distribution generated manually in Table 4.4.

4.3 Percentiles and Percentile Ranks

4.3.1 Percentiles

Percentiles (*P*) are measures of relative standing and it is used primarily to compare the performance of an individual against that of a reference group. Let's take your IQ score of 120 as an example. The IQ score value of 120 is simply a number and in and of itself, it is meaningless. It takes on meaning only when it can be compared with some standard score or base value such as the percentage of people with lower IQ scores than yours. Thus, if you inform your family that 65% of the IQ scores from your class of 100 students were below yours, then your score of 120 would be the 65th percentile, that is, you have scored higher than 65% of your classmates.

From the above example, it can be seen that a percentile is a measure indicating the value below which a given percentage of observations in a group of observations fall. Thus, the 30th percentile is the value (or score) below which 30% of the observations fall.

4.3.2 Computation of Percentiles (Finding the Score below which a Specified Percentage of Scores will Fall)

Suppose that one of the 100 students who took the IQ test informs you that his IQ score was at the 50th percentile (P_{50}). What was his IQ score? To determine the score that corresponds to the 50th percentile (P_{50}), we will use the grouped frequency distribution found in Table 4.5. Perhaps the easiest way to calculate any percentile is to apply the following equation (Pagano, 2013):

$$Percentile = X_L + (i/f_i)(cum\ f_P - cum\ f_L)$$

where

X_L = value of the lower real limit of the interval containing the percentile

$cum\ f_P$ = frequency of scores below the percentile

$cum\ f_L$ = frequency of scores below the lower real limit of the interval containing the percentile

f_i = frequency of the interval containing the percentile

i = interval width

For the present example (see Table 4.5),

$$X_L = 108.5$$
$$cum\ f_P = 50$$
$$cum\ f_L = 32$$
$$f_i = 23$$
$$i = 7$$

Substituting these values in the equation, we calculate P_{50} (the 50th percentile) as

$$\text{Percentile} = X_L + (i/f_i)(cum\ f_P - cum\ f_L)$$

$$P_{50} = 108.5 + (7/23)(50 - 32)$$
$$P_{50} = 108.5 + (0.3043)(18)$$
$$P_{50} = 108.5 + 5.4774$$
$$P_{50} = \mathbf{113.9774}$$

4.3.3 SPSS Syntax Method

We can calculate P_{50} (the 50th percentile) for the above example using the SPSS syntax method (SPSS does not provide the Windows method for calculating percentiles from grouped frequency distributions). The first thing that needs to be done is to set up a data file that contains the variables and their values listed in the equation.

4.3.4 Data Entry Format

The data set has been saved under the name: **EX2.SAV**

Variables	Value
X_L	108.5
$cum\ f_P$	50
$cum\ f_L$	32
f_i	23
i	7

4.3.5 SPSS Syntax Method

```
COMPUTE P50=XL+((i/fi)*(cum_fP-cum_fL)).
EXECUTE.
```

1. From the menu bar, click **File**, then **New**, and then **Syntax.** The following **IBM SPSS Statistics Syntax Editor** window will open.

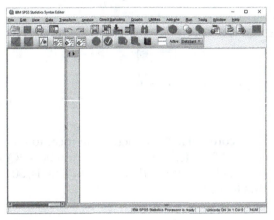

2. Type the **Compute** syntax command in the **IBM SPSS Statistics Syntax Editor** window.

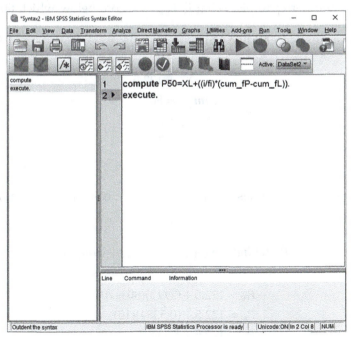

3. To run the **Compute** command, click ▶ or click Run and then **All**.

4.3.6 SPSS Output

Successful execution of the **Compute** command will yield the P_{50} (50th percentile) score, which is added to the **EX2.SAV** data file under the variable name **P50**. The following **Data View** screen shows the IQ score corresponding to P_{50} (50th percentile).

		XL	cum_fP	cum_fL	fi	i		P50
1		108.50	50.00	32.00	23.00	7.00		113.98

"EX2.sav [DataSet2] - IBM SPSS Statistics Data Editor
File   Edit   View   Data   Transform   Analyze   Direct Marketing   Graphs   Utilities   Add-ons   Window   Help

The computed P_{50} score is **113.98**, which is identical to the value calculated manually (see Section 4.3.2). Thus, for the student whose IQ score was at the 50th percentile (P_{50}), his IQ score would be 113.98. That is, 50% of the observations fall below this IQ score.

4.3.7 Another Example

Let us try another example. What is P_{80}, the score below which 80% of the scores fall?

For this example (see Table 4.5),

$$X_L = 122.5$$
$$cum\, f_P = 80$$
$$cum\, f_L = 70$$
$$f_i = 13$$
$$i = 7$$

Substituting these values in the equation, we calculate P_{80} (the 80th percentile) as

$$Percentile = X_L + (i/f_i)(cum\, f_P - cum\, f_L)$$

$$P_{80} = 122.5 + (7/13)(80 - 70)$$
$$P_{80} = 122.5 + (0.5384)(10)$$
$$P_{80} = 122.5 + 5.384$$
$$P_{80} = \mathbf{127.884}$$

We can calculate P_{80} (the 80th percentile) for the above example using the SPSS syntax method. Set up a data file that contains the variables and their values listed in the equation.

4.3.8 Data Entry Format

The data set has been saved under the name: **EX3.SAV**

Variables	Value
X_L	122.5
$cum\ f_P$	80
$cum\ f_L$	70
f_i	13
I	7

4.3.9 SPSS Syntax Method

```
COMPUTE P80=XL+((i/fi)*(cum_fP-cum_fL)).
EXECUTE.
```

1. From the menu bar, click **File**, then **New**, and then **Syntax.** The following **IBM SPSS Statistics Syntax Editor** window will open.

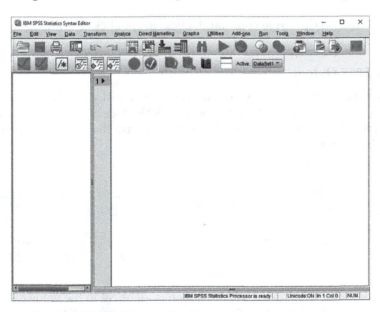

2. Type the **Compute** syntax command in the **IBM SPSS Statistics Syntax Editor** window.

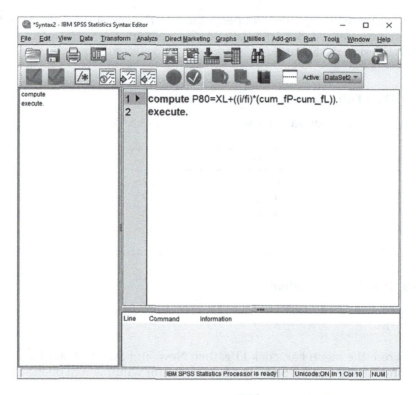

3. To run the **Compute** command, click ▶ or click Run and then **All**.

4.3.10 SPSS Output

Successful execution of the **Compute** command will yield the P_{80} (80th percentile) score, which is added to the **EX3.SAV** data file under the variable name **P80**. The following **Data View** screen shows the IQ score corresponding to P_{80} (80th percentile).

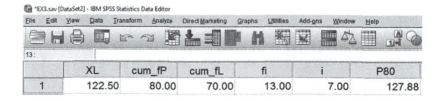

The computed P_{80} score is **127.88**, which is identical to the value calculated manually (see Section 4.3.7). Thus, for the student who's IQ score was at the 80th percentile (P_{80}), his IQ score would be 127.88. That is, 80% of the observations fall below this score.

4.3.11 Percentile Rank

A *percentile rank (PR)* is simply the reverse of calculating a percentile. Whereas a percentile is concerned with the calculation of the score below which a specified percentage of scores fall, a percentile rank is concerned with the calculation of the percentage of scores that fall below a given score. That is, given a particular score, what is the percentage of scores that fall below this score?

4.3.12 Computation of Percentile Ranks (Finding the Percentage of Scores that Fall below a Given Score)

Suppose that one of the 100 students who took the IQ test informs you that his IQ score was 127. What is the percentage of scores that fall below this score (i.e., the percentile rank)? As with the calculation of percentiles, we will use the grouped frequency distribution found in Table 4.5 to solve this problem. To calculate any percentile rank, we will apply the following equation (Pagano, 2010):

$$\text{Percentile rank } (PR_{127}) = \frac{cum\, f_L + \left(\dfrac{f_i}{i}\right)(X - XL)}{N} \times 100$$

where

$cum\, f_L$ = frequency of scores below the lower real limit of the interval containing the score X

X = score whose percentile rank is being determined

X_L = value of the lower real limit of the interval containing the score X

i = interval width

f_i = frequency of the interval containing the score X

N = total number of raw scores

For the present example (see Table 4.5),

$$cum\, f_L = 70$$
$$X = 127$$
$$XL = 122.5$$
$$i = 7$$
$$f_i = 13$$
$$N = 100$$

Substituting these values in the equation we calculate the percentile rank for the IQ score of 127 as

$$\text{Percentile rank} = \frac{cum\,f_L + \left(\dfrac{f_i}{i}\right)(X - X_L)}{N} \times 100$$

$$\text{Percentile rank}\,(PR_{127}) = \frac{70 + \left(\dfrac{13}{7}\right)(127 - 122.5)}{100} \times 100$$

$$\text{Percentile rank}\,(PR_{127}) = \frac{70 + (1.8571)(4.5)}{100} \times 100$$

$$\text{Percentile rank}\,(PR_{127}) = \frac{70 + (8.3569)}{100} \times 100$$

$$\text{Percentile rank}\,(PR_{127}) = \mathbf{78.36}$$

We can calculate percentile rank for the IQ score of 127 for the above example using the SPSS syntax method. The first thing that needs to be done is to set up a data file that contains the variables and their values listed in the equation.

4.3.13 Data Entry Format

The data set has been saved under the name: **EX4.SAV**

Variables	Value
$cum\,f_L$	70
X	127
XL	122.5
i	7
f_i	13
N	100

4.3.14 SPSS Syntax Method

```
COMPUTE PR127=(cum_fL+((fi/i)*(X-XL)))/N*100.
EXECUTE.
```

1. From the menu bar, click **File**, then **New**, and then **Syntax.** The following **IBM SPSS Statistics Syntax Editor** window will open.

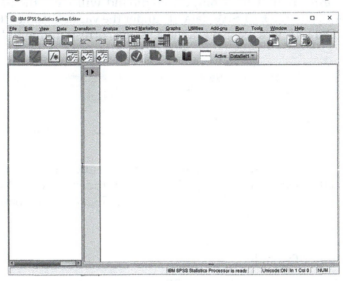

2. Type the **Compute** syntax command in the **IBM SPSS Statistics Syntax Editor** window.

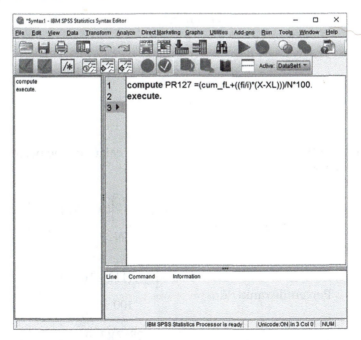

3. To run the **Compute** command, click ▶ or click Run and then **All**.

4.3.15 SPSS Output

Successful execution of the **Compute** command will yield the PR_{127} score, which is added to the **EX4.SAV** data file under the variable name **PR127**. The following **Data View** screen shows the IQ score corresponding to PR_{127}.

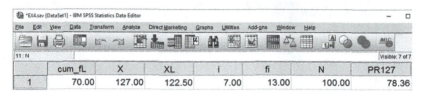

	cum_fL	X	XL	i	fi	N	PR127
1	70.00	127.00	122.50	7.00	13.00	100.00	78.36

The computed PR_{127} score is **78.36**, which is identical to the value calculated manually (see Section 4.3.12). Thus, for the student whose IQ score was 127, his percentile rank would be 78.36. That is, 78.36% of the observations fall below this IQ score of 127.

4.3.16 Another Example

Let's try another example. What is PR_{112}, the percentage of scores that fall below this IQ score of 112?

For this example (see Table 4.5),

$$cum\ f_L = 32$$
$$X = 112$$
$$X_L = 108.5$$
$$i = 7$$
$$f_i = 23$$
$$N = 100$$

Substituting these values in the equation, we calculate the percentile rank for the IQ score of 112 as

$$\text{Percentile rank} = \frac{cum\ f_L + \left(\dfrac{f_i}{i}\right)(X - X_L)}{N} \times 100$$

$$\text{Percentile rank}\,(PR_{112}) = \frac{32 + \left(\dfrac{23}{7}\right)(112 - 108.5)}{100} \times 100$$

$$\text{Percentile rank}\,(PR_{112}) = \frac{32 + (3.2857)(3.5)}{100} \times 100$$

$$\text{Percentile rank } (PR_{112}) = \frac{32 + (11.4999)}{100} \times 100$$

Percentile rank $(PR_{112}) = \textbf{43.4999 (43.5)}$

We can calculate PR_{112} for the above example using the SPSS syntax method. Set up a data file that contains the variables and their values listed in the equation.

4.3.17 Data Entry Format

The data set has been saved under the name: **EX5.SAV**

Variables	Value
$cum\,f_L$	32
X	112
X_L	108.5
i	7
f_i	23
N	100

4.3.18 SPSS Syntax Method

```
COMPUTE PR112 = (cum_fL + ((fi/i) * (X-XL))) /N*100.
EXECUTE.
```

1. From the menu bar, click **File**, then **New**, and then **Syntax.** The following **IBM SPSS Statistics Syntax Editor** window will open.

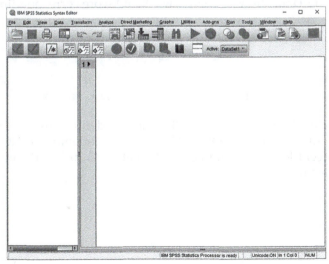

2. Type the **Compute** syntax command in the **IBM SPSS Statistics Syntax Editor** window.

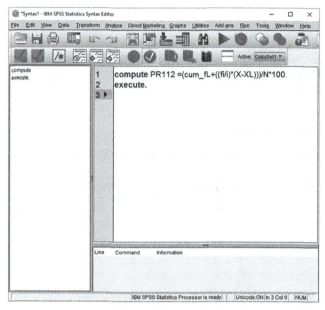

3. To run the **Compute** command, click ▶ or click Run and then **All**.

4.3.19 SPSS Output

Successful execution of the **Compute** command will yield the PR_{112} score, which is added to the **EX5.SAV** data file under the variable name **PR112**. The following **Data View** screen shows the IQ score corresponding to PR_{112}.

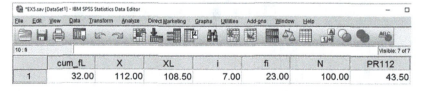

The computed PR_{112} score is **43.50**, which is identical to the value calculated manually (see Section 4.3.16). Thus, for the student whose IQ score was 112, his percentile rank would be 43.50. That is, 43.50% of the observations fall below his IQ score of 112.

5

Graphing

5.1 Graphing Frequency Distributions

Frequency distributions are often presented in graphical form because graphs make information easier to visualize. Using visual representations to present data from surveys or other sources make them easier to understand. This is because graphs condense large amounts of information into easy-to-understand formats that clearly and effectively communicate important points. *Bar graphs, histograms, frequency polygons,* and *cumulative percentage curves* are excellent ways to illustrate your research results.

5.2 Bar Graph

Frequency distributions of *nominal/categorical* or *ordinal* data are often graphed using bar graphs. A bar graph comprises discrete bars that represent different categories of data (e.g., **gender**: male, female; **ethnicity**: Chinese, Thai, Australian). The height of each bar is equal to the quantity within that category of data. Bar graphs are used most often to compare values across categories. Please note that for categorical/nominal data, there is no quantitative relationship between the categories and this characteristic is emphasized by the fact that the bars in a bar graph do not touch each other, that is, there are gaps between them. The following illustrates how to draw a bar graph using the SPSS Windows method and the SPSS Syntax method.

5.2.1 An Example

Suppose that in a survey study, the following question was asked:

What is your current employment status?

1. Full-time employment
2. Part-time employment

3. Household duties

4. Unemployed

5. Student studying only

6. Student working part-time

In this example, the respondent is asked to select the employment category that best describes his/her employment status. These six employment categories are subsumed under the variable **EMPLOY**.

5.2.2 Data Entry Format

The data set has been saved under the name: **EX6.SAV**

Variables	Column(s)	Code
SEX	1	1 = male, 2 = female
AGE	2–3	In years
EDUC	4	1 = Primary
		2 = 1–2 years secondary
		3 = 3–4 years secondary
		4 = 5–6 years secondary
		5 = Technical/trade
		6 = Tertiary
EMPLOY	5	1 = Full-time employment
		2 = Part-time employment
		3 = Household duties
		4 = Unemployed
		5 = Student studying only
		6 = Student working part-time
Marital	6	1 = Never married
		2 = Married
		3 = Separated
		4 = Divorced
		5 = De facto
		6 = Widowed
PARTY	7	1 = Labor
		2 = Liberal
		3 = National
		4 = Democrat

5.2.3 SPSS Windows Method

1. In order to draw a bar graph of the frequency distribution representing the **EMPLOY** variable, open the data set **EX6.SAV**. Click **Graphs** on the menu bar, then **Legacy Dialogs**, and then **Bar**. The following **Bar Charts** Window will open.

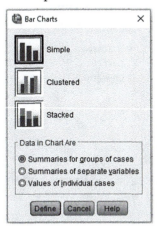

2. Click (highlight) the icon. Make sure that the **Summaries for Groups of Cases** bullet-point is checked. Click **Define** to open the **Define Simple Bar: Summaries for Groups of Cases** Window below.

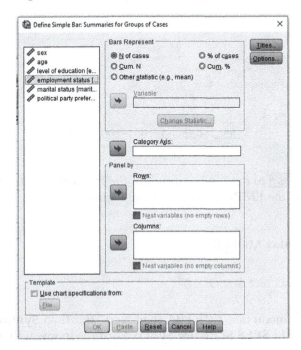

3. When the **Define Simple Bar: Summaries for Groups of Cases** Window opens, check the bullet-point **N of cases**. Next, transfer the **EMPLOY** variable to the **Category axis:** cell by clicking the **EMPLOY** variable (highlight) and then 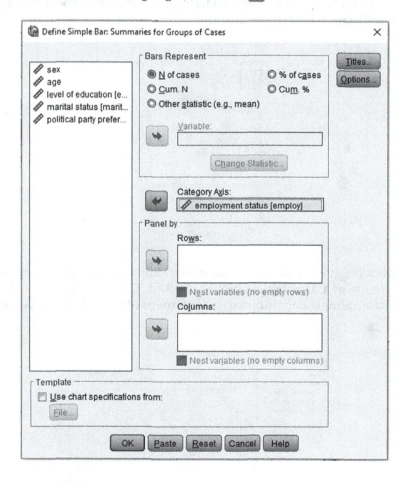.

Click [OK] to draw a bar graph of the frequency distribution representing the **EMPLOY** variable (see Figure 5.1).

5.2.4 SPSS Syntax Method

```
GRAPH
/BAR(SIMPLE)=COUNT BY EMPLOY.
```

1. From the menu bar, click **File**, then **New**, and then **Syntax.** The following **IBM SPSS Statistics Syntax Editor** Window will open.

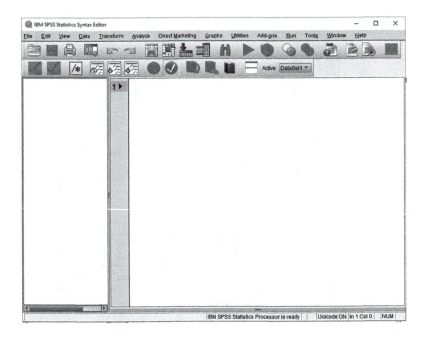

2. Type the **Bar Graph** syntax command in the **IBM SPSS Statistics Syntax Editor** Window.

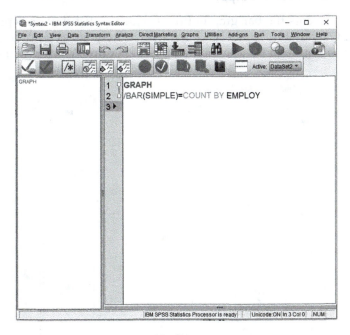

3. To run the **Bar Graph** analysis, click ▶ or click Run and then **All**.

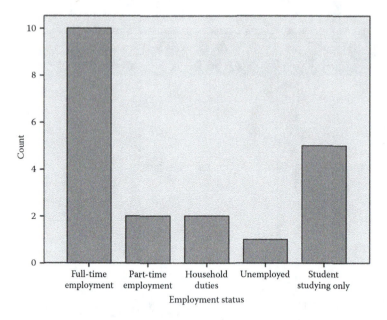

FIGURE 5.1
Bar graph of the frequency distribution representing the EMPLOY variable.

5.2.5 SPSS Bar Graph Output

Figure 5.1 presents the bar graph of the frequency distribution representing the **EMPLOY** variable. To label each bar with its frequency count, double-click on Figure 5.1 within SPSS. The following **Chart Editor** Window will open.

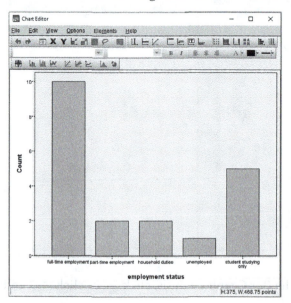

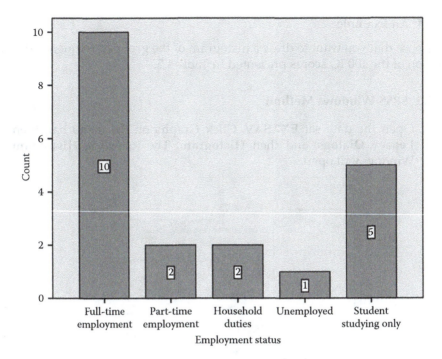

FIGURE 5.2
Bar graph and the frequency counts of the distribution representing the EMPLOY variable.

Next, click on the **Show Data Labels** icon . This procedure will label each bar in the bar graph with its frequency count. Close this Window and the bar graph of the frequency distribution representing the **EMPLOY** variable will include the frequency count for each bar (see Figure 5.2).

From Figure 5.2, it can be seen that the majority of the respondents are in full-time employment ($N = 10$; 50%).

5.3 Histogram

Frequency distributions of *interval* or *ratio* data are often graphed using histograms. Like a bar chart, a histogram is made up of columns plotted on a graph with the height of the column representing the frequency count of that group defined by the column label. The main difference between bar charts and histograms is that with bar charts, each column represents a group (e.g., male, female) defined by a categorical variable (gender). With histograms, each column represents a class interval defined by a quantitative variable (e.g., age, weight, height, class interval).

5.3.1 An Example

Suppose that you want to draw a histogram of the grouped frequency distribution of the 100 IQ scores presented in Table 4.5.

5.3.2 SPSS Windows Method

1. Open the data set **EX7.SAV.** Click **Graphs** on the menu bar, then **Legacy Dialogs**, and then **Histogram**. The following **Histogram** Window will open.

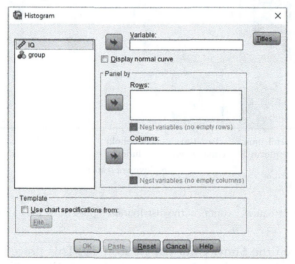

2. Transfer the **GROUP** variable to the **Variable:** cell by clicking the **GROUP** variable (highlight) and then 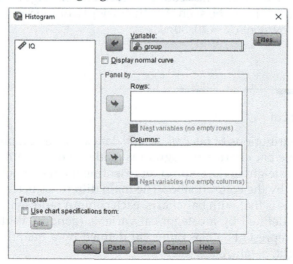.

Click [OK] to draw a histogram of the frequency distribution representing the **GROUP** variable (see Figure 5.3).

5.3.3 SPSS Syntax Method

```
GRAPH
/HISTOGRAM=GROUP.
```

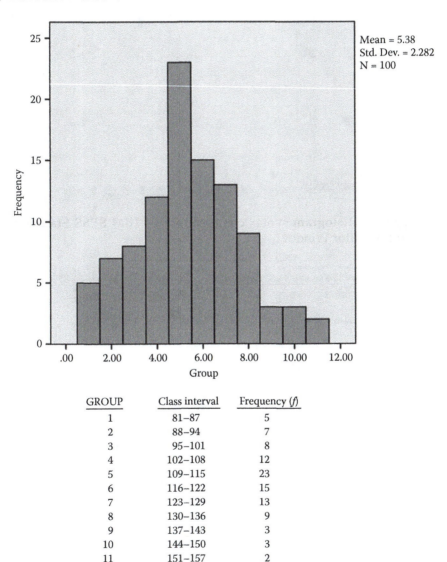

GROUP	Class interval	Frequency (*f*)
1	81–87	5
2	88–94	7
3	95–101	8
4	102–108	12
5	109–115	23
6	116–122	15
7	123–129	13
8	130–136	9
9	137–143	3
10	144–150	3
11	151–157	2

FIGURE 5.3
Histogram of the frequency distribution representing the GROUP variable.

1. From the menu bar, click **File**, then **New**, and then **Syntax**. The following IBM SPSS **Statistics Syntax Editor** Window will open.

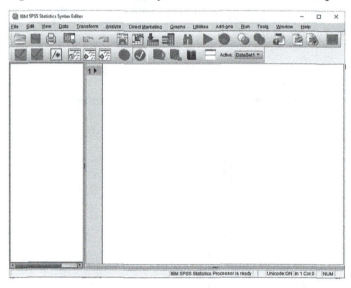

2. Type the **Histogram** syntax command in the **IBM SPSS Statistics Syntax Editor** Window.

3. To run the **Histogram** analysis, click ▶ or click Run and then **All**.

5.3.4 SPSS Histogram Output

Figure 5.3 presents the histogram of the frequency distribution representing the **GROUP** variable. To label each bar with its frequency count, double-click on Figure 5.3 within SPSS. The following **Chart Editor** Window will open.

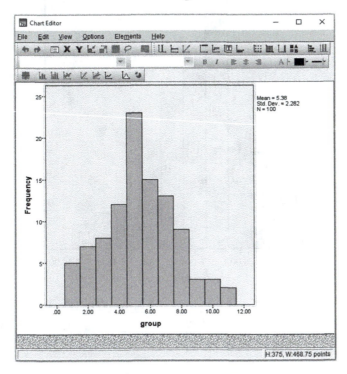

Next, click on the **Show Data Labels** icon. This procedure will label each bar in the histogram with its frequency count. Close this Window and the histogram of the frequency distribution representing the **GROUP** variable will include the frequency count for each bar (see Figure 5.4).

From Figure 5.4, it can be seen that the majority of the IQ scores ($N = 63$) are between the class intervals of 102–108 (**GROUP** = 4) and 123–129 (**GROUP** = 7).

5.4 Frequency Polygon

Frequency distributions of *interval* or *ratio* data can also be represented by frequency polygons. Like a histogram, the horizontal axis represents the class intervals defined by a quantitative variable (e.g., age, weight, height, class interval). The main difference between histograms and frequency

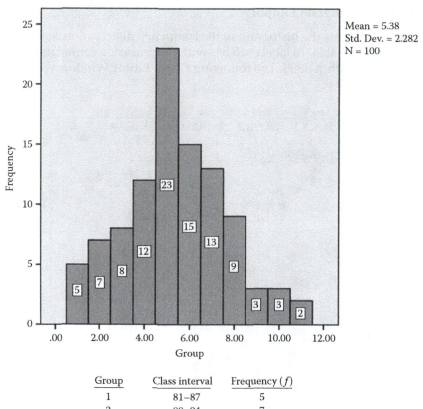

Group	Class interval	Frequency (f)
1	81–87	5
2	88–94	7
3	95–101	8
4	102–108	12
5	109–115	23
6	116–122	15
7	123–129	13
8	130–136	9
9	137–143	3
10	144–150	3
11	151–157	2

FIGURE 5.4
Histogram and the frequency counts of the distribution representing the GROUP variable.

polygons is that the bars of the histogram are replaced by points plotted over the midpoints of the class intervals at a height corresponding to the frequencies of the class intervals.

5.4.1 An Example

Suppose you want to draw a frequency polygon of the grouped frequency distribution of the 100 IQ scores presented in Table 4.5.

5.4.2 SPSS Windows Method

1. Open the data set **EX7.SAV.** Click **Graphs** on the menu bar, then **Legacy Dialogs**, and then **Line**. The following **Line Charts** Window will open.

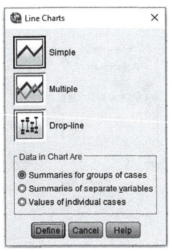

2. Click (highlight) the 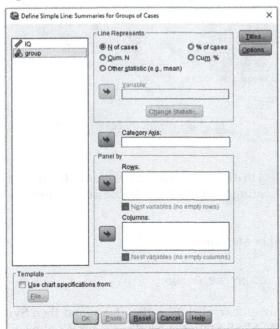 icon. Make sure that the **Summaries for groups of cases** bullet-point is checked. Click [Define] to open the **Define Simple Bar: Summaries for Groups of Cases** Window below.

3. When the **Define Simple Line: Summaries for Groups of Cases** Window opens, check the bullet-point <u>N</u> **of cases**. Next, transfer the **GROUP** variable to the **Category a<u>x</u>is:** cell by clicking the **GROUP** variable (highlight) and then 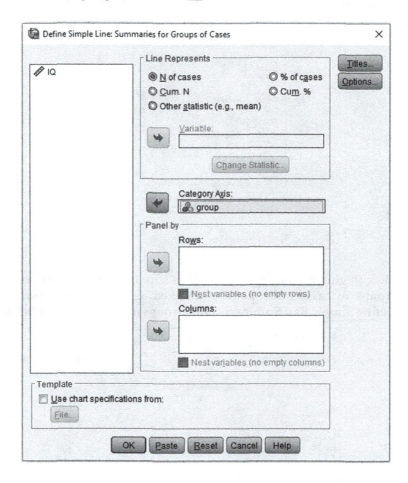.

Click [OK] to draw a frequency polygon of the frequency distribution representing the **GROUP** variable (see Figure 5.5).

5.4.3 SPSS Syntax Method

```
GRAPH
/LINE(SIMPLE)=COUNT BY GROUP.
```

1. From the menu bar, click **File**, then **New**, and then **Syntax.** The following **IBM SPSS Statistics Syntax Editor** Window will open.

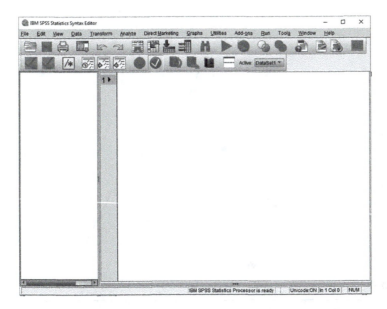

2. Type the **Frequency polygon** syntax command in the **IBM SPSS Statistics Syntax Editor** Window.

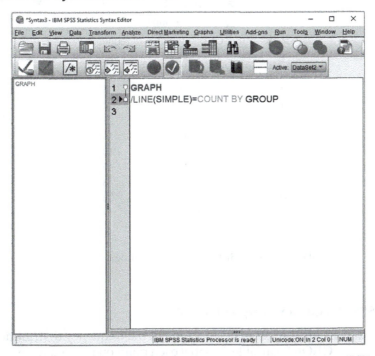

3. To run the **Frequency polygon** analysis, click ▶ or click Run and then **All**.

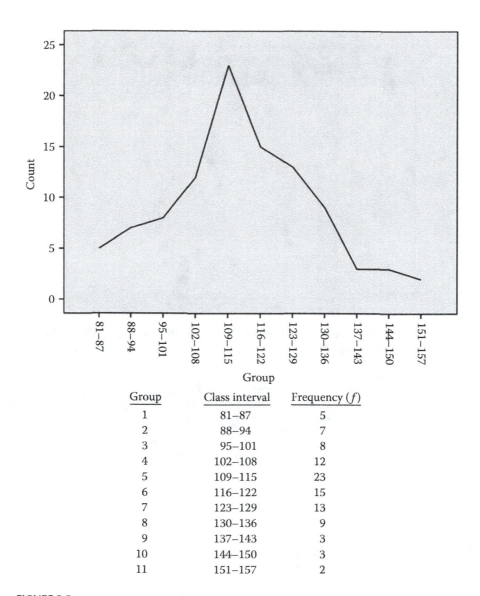

Group	Class interval	Frequency (f)
1	81–87	5
2	88–94	7
3	95–101	8
4	102–108	12
5	109–115	23
6	116–122	15
7	123–129	13
8	130–136	9
9	137–143	3
10	144–150	3
11	151–157	2

FIGURE 5.5
Frequency polygon of the frequency distribution representing the GROUP variable.

5.4.4 SPSS Frequency Polygon Output

Figure 5.5 presents the frequency polygon of the frequency distribution representing the **GROUP** variable. To label each point on the frequency polygon with its frequency count, double-click on Figure 5.5 within SPSS. The following **Chart Editor** Window will open.

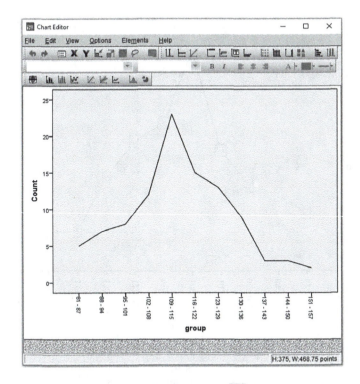

Next, click on the **Show Data Labels** icon [image]. This procedure will label each point on the frequency polygon with its frequency count. Close this Window and the frequency polygon of the frequency distribution representing the **GROUP** variable will include the frequency count for each point (see Figure 5.6).

It can be seen that the frequency polygon results (see Figure 5.6) are similar to the histogram output results (see Figure 5.4) in that the majority of the IQ scores ($N = 63$) are within the class intervals of 102–108 (**GROUP** = 4) and 123–129 (**GROUP** = 7).

5.5 Cumulative Percentage Curve

Cumulative percentage curve, also called an *ogive*, is another way of expressing frequency distribution by calculating the percentage of the cumulative frequency within each interval. It should be noted that cumulative frequency and cumulative percentage graphs are exactly the same. The main advantage of cumulative percentage over cumulative frequency as a measure of frequency distribution is that it provides an easier way to compare different sets of data. This is because the vertical axis scale of the cumulative percentage graph is scaled in percentages (0%–100%), which allows for direct comparisons

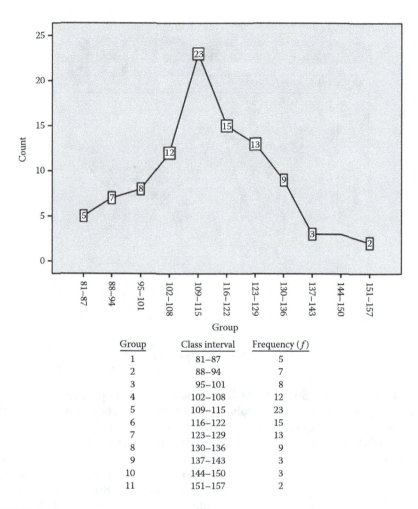

Group	Class interval	Frequency (f)
1	81–87	5
2	88–94	7
3	95–101	8
4	102–108	12
5	109–115	23
6	116–122	15
7	123–129	13
8	130–136	9
9	137–143	3
10	144–150	3
11	151–157	2

FIGURE 5.6

Frequency polygon and the frequency counts of the distribution representing the GROUP variable.

between different data sets. Cumulative percentage is calculated by (1) dividing the cumulative frequency by the total number of observations (n) and (2) multiplying it by 100 (the last value will always be equal to 100%). Thus,

$$\textbf{Cumulative percentage} = \left(\frac{\textbf{cumulative frequency}}{n}\right) \times \textbf{100}$$

5.5.1 An Example

Suppose you want to draw a cumulative percentage curve of the frequency distribution of the 100 IQ scores presented in Table 4.2.

5.5.2 SPSS Windows Method

1. Open the data set **EX7.SAV.** Click **Graphs** on the menu bar, then **Legacy Dialogs**, and then **Line**. The following **Line Charts** Window will open.

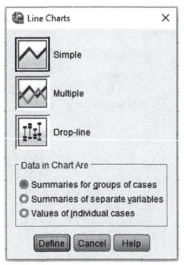

2. Click (highlight) the 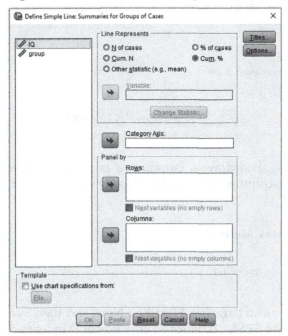 icon. Make sure that the **Summaries for Groups of Cases** bullet-point is checked. Click **Define** to open the **Define Simple Bar: Summaries for Groups of Cases** Window below.

3. When the **Define Simple Line: Summaries for Groups of Cases** Window opens, check the bullet-point **Cum %**. Next, transfer the **IQ** variable to the **Category axis:** cell by clicking the **IQ** variable (highlight) and then ▶.

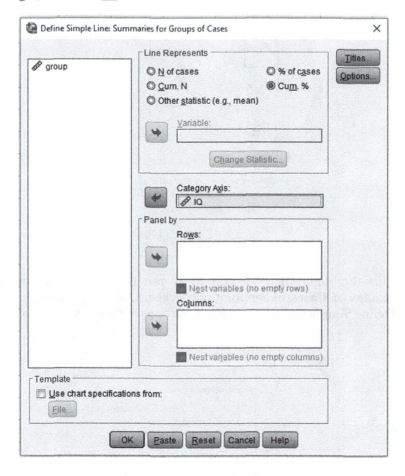

Click [OK] to draw a cumulative percentage curve of the frequency distribution representing the **IQ** variable (see Figure 5.7).

5.5.3 SPSS Syntax Method

```
GRAPH
/LINE(SIMPLE)=CUPCT BY IQ.
```

1. From the menu bar, click **File**, then **New**, and then **Syntax.** The following **IBM SPSS Statistics Syntax Editor** Window will open.

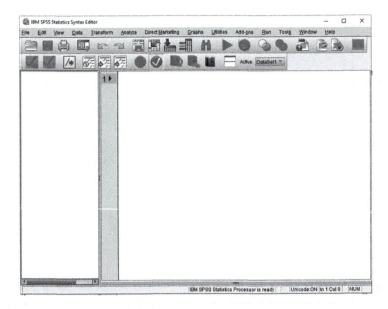

2. Type the **cumulative percentage** syntax command in the **IBM SPSS Statistics Syntax Editor** Window.

3. To run the **cumulative percentage** analysis, click ▶ or click Run and then **All**.

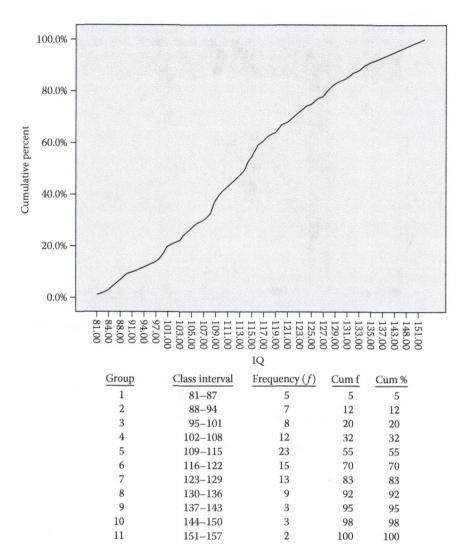

Group	Class interval	Frequency (f)	Cum f	Cum %
1	81–87	5	5	5
2	88–94	7	12	12
3	95–101	8	20	20
4	102–108	12	32	32
5	109–115	23	55	55
6	116–122	15	70	70
7	123–129	13	83	83
8	130–136	9	92	92
9	137–143	3	95	95
10	144–150	3	98	98
11	151–157	2	100	100

FIGURE 5.7
Cumulative percentage curve of the frequency distribution representing the 100 IQ scores.

5.5.4 SPSS Cumulative Percentage Output

Figure 5.7 presents the cumulative percentage curve of the frequency distribution representing the 100 IQ scores (from Table 4.2). Both *percentiles* and *percentile ranks* can be read directly off the cumulative percentage curve. For example, what is the 50th percentile (P_{50}) for this data set of 100 IQ scores? (i.e., what is the IQ score below which 50% of the scores will fall?). To find P_{50}, retrieve Figure 5.7 and then on the Y-axis draw a horizontal line from the scaled value of 50% to where the line intersects the cumulative percentage

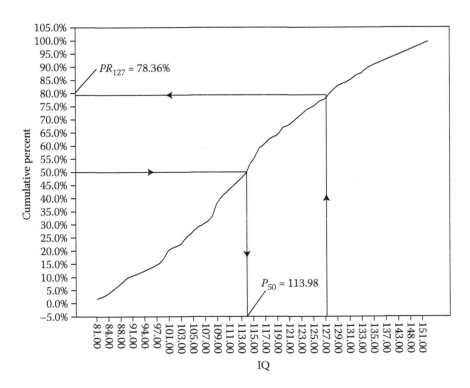

FIGURE 5.8
Cumulative percentage curve: IQ scores from Figure 5.7.

curve. From this intersection point, draw a vertical line downwards until it touches the X-axis. Where the line touches the X-axis is the IQ score value that corresponds to P_{50} (**113.98**) (see Figure 5.8).

The procedure for reading percentile ranks directly off the cumulative percentage curve is simply the opposite of the procedure for reading percentiles described above. For example, what is the percentile rank for the IQ score of 127 (PR_{127})? (i.e., what percentage of IQ scores fall below the IQ score of 127?). To find PR_{127}, retrieve Figure 5.7 and then on the X-axis draw a vertical line upwards from the IQ score of 127 to where the line intersects the cumulative percentage curve.

From this intersection point, draw a horizontal line until it touches the Y-axis. Where the line touches the Y-axis is the percentage of IQ scores that fall below the IQ score of 127 (**78.36%**) (see Figure 5.8).

6

Measures of Central Tendency

6.1 Why Is Central Tendency Important?

Central tendency is very useful in the social sciences. It allows the researcher to know what the *average* for a data set is. It also summarizes the data into one quantitative value, which can be used to represent the entire data set. For example, from Chapter 4 we noted that the primary function of frequency distributions is to organize and present data in meaningful ways. The frequency distribution of the 100 IQ scores presented in Table 4.2 organized the data into an easily interpretable form – that is, the 100 IQ scores are rank-ordered and the scores' frequency counts, percent, and cumulative percent are computed. However, as useful as this information is, the frequency distribution does not, in itself, allows us to make quantitative statements that characterize the distribution as a whole. For example, looking at the frequency distribution of the 100 IQ scores presented in Table 4.2, it can be seen that some of the score values are high, some are low, and some are moderate. But neither do these scores allow us to summarize or to characterize the distribution quantitatively nor do they allow for quantitative comparisons to be made between two or more distributions.

For example, a psychologist may be interested in investigating whether there is any gender difference in IQ scores. To achieve this aim, the IQ scores from a sample of 10 males and 10 females is collected. Table 6.1 presents the distribution of IQ scores from this group of respondents.

Looking at the frequency distribution of IQ scores for the male and female respondents, it is very difficult to make any quantitative statements about possible difference in their IQ. This is because some of the scores are higher for males than for females and vice versa. The question then becomes how can the psychologist compare the two distributions? This is most often done by calculating the average IQ for each group and then comparing the averages. For this example, the computed average IQ score for males is **103.5** and for females it is **102.3**. Thus, calculating the groups' averages and comparing them allow the psychologist to make the quantitative statement that there is gender difference in IQ scores and that the

TABLE 6.1

Distribution of IQ Scores from 10 Male
and 10 Female Respondents

Respondents	Male	Female
1	110	115
2	106	110
3	120	108
4	96	125
5	130	105
7	105	112
8	115	98
9	118	120
10	135	130

male respondents scored higher on average than the female respondents. From this example, it can be seen that measures of central tendency greatly simplify the task of drawing conclusions.

6.2 Measures of Central Tendency

A measure of central tendency is a *summary measure that describes an entire set of data with a single value that represents the middle or center of its distribution.*

There are three main measures of central tendency: the **mean**, the **median**, and the **mode**. While they are all valid measures of central tendency, some become more appropriate to use than others under different conditions. In the following sections, we will look at these three measures of central tendency and learn how to calculate them and understand the conditions under which they are most appropriate to be used.

6.3 The Arithmetic Mean

Most people are familiar with the *arithmetic mean*. It is the average value for a variable that is calculated by summing the scores of that variable and dividing the summed value by the number of scores. For example, from Table 6.1, we calculate the mean IQ score for the 10 male respondents by summing the

10 'male' IQ scores and dividing the summed value by 10. In short, the arithmetic mean is defined as *the sum of the scores divided by the number of scores.* Stated in algebraic form:

$$\bar{x} = \frac{\sum X}{N} = \frac{X_1 + X_2 + \cdots + X_N}{N} = \text{sample mean}$$

$$\mu = \frac{\sum X}{N} = \frac{X_1 + X_2 + \cdots + X_N}{N} = \text{population mean}$$

where
$\bar{x}$ = mean of a set of sample scores
μ (mu) = mean of a set of population scores
$\sum$ = summation (sigma)
X_1 to X_N = list of raw scores
N = number of scores

6.3.1 How to Calculate the Arithmetic Mean

Let's say we want to calculate the mean for the set of 100 IQ scores presented in Table 4.2.

6.3.2 SPSS Window Method

1. Launch the SPSS program and then open the data file **EX1.SAV.** Click **Analyze** on the menu bar, then **Descriptive Statistics,** and then **Frequencies.** The following **Frequencies** Window will open.

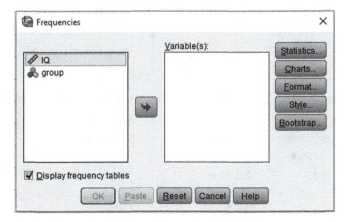

2. In the left-hand field containing the study's **IQ** variable, click (high-light) this variable, and then click 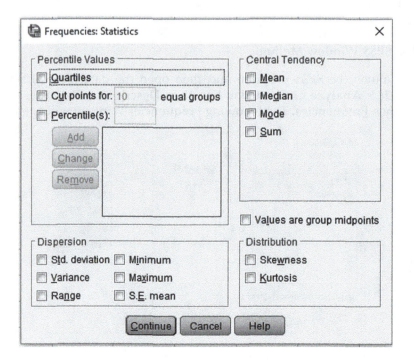 to transfer the selected IQ variable to the **Variable(s):** field.

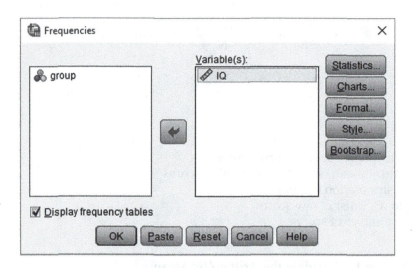

3. Click to open the **Frequencies Statistics** Window below.

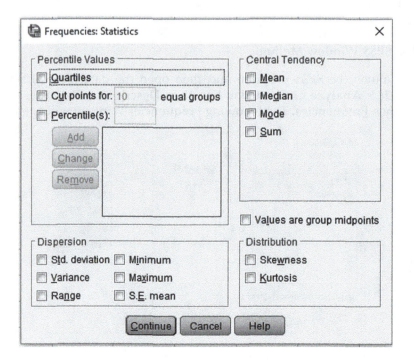

4. Under the **Central Tendency** heading, check the <u>M</u>ean cell.

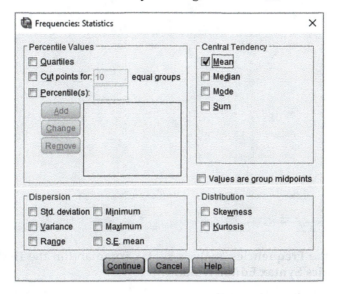

Click [Continue] to return to the **Frequencies** Window.

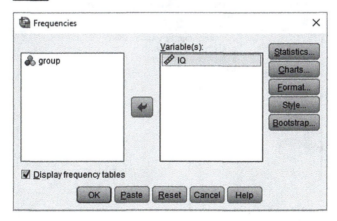

5. Click [OK] to run the analysis. See Table 6.2 for the results.

6.3.3 SPSS Syntax Method

```
FREQUENCIES VARIABLES=IQ
/STATISTICS=MEAN
/ORDER=ANALYSIS.
```

1. From the menu bar, click **File**, then **New**, and then **Syntax.** The following **IBM SPSS Statistics Syntax Editor** Window will open.

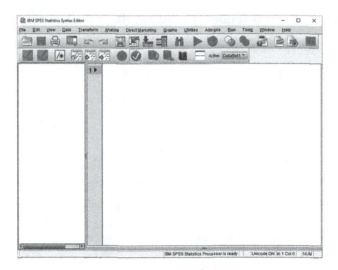

2. Type the **Frequencies** analysis syntax command in the **IBM SPSS Statistics Syntax Editor** Window.

3. To run the **Frequencies** analysis, click ▶ or click **Run** and then **All**.

6.3.4 SPSS Output

As can be seen from Table 6.2, the mean IQ score for the set of 100 IQ scores (see Table 4.2) is **114.54**. That is, the average IQ score for the sample of 100 respondents is 114.54.

TABLE 6.2

Frequencies Output of Mean IQ Score

Statistics		
IQ		
N	Valid	100
	Missing	0
Mean		114.5400

6.3.5 How to Calculate the Mean from a Grouped Frequency Distribution

The procedures for finding the mean from a grouped frequency distribution are similar to the procedures that are employed with the ungrouped frequency distributions. The procedures involve the following steps.

Step 1: Find the midpoint (*x*) of each interval. The midpoint of a class interval is calculated as

Midpoint of interval $(x) = 0.5 \times$ **(Lower class limit + Upper class limit)**

Thus, the midpoint for an interval 15–19 is $0.5 \times (15 + 19) = $ **17**

Step 2: Multiply the frequency (*f*) of each interval by its midpoint (i.e., *fx*).

Step 3: Get the sum of all the frequencies (i.e., *N*) and the sum of all the *fx* (i.e., Σfx). Divide Σfx by *N* to get the mean, that is, $\bar{x} = \dfrac{\Sigma fx}{N}$

6.3.6 An Example

Suppose you want to calculate the mean from the grouped frequency distribution of the 100 IQ scores presented in Table 4.4. The table of scores is presented here for your convenience.

Class Interval	Frequency (*f*)	Midpoint (*x*)	Frequency X Midpoint (*fx*)
151–157	2	154	308
144–150	3	147	441
137–143	3	140	420
130–136	9	133	1197
123–129	13	126	1638
116–122	15	119	1785
109–115	23	112	2576
102–108	12	105	1260
95–101	8	98	784
88–94	7	91	637
81–87	5	84	420
	N = 100		Σfx = 11466

Thus,

$$\bar{x} = \frac{\sum fx}{N}$$

$$\bar{x} = \frac{11466}{100}$$

$$\bar{x} = 114.66$$

The mean of the grouped frequency distribution of the 100 IQ scores is **114.66.**

6.3.7 Calculating the Mean from Grouped Frequency Distribution Using SPSS

The above example can be demonstrated using the SPSS syntax method (SPSS does not provide the Windows method for calculating means from grouped frequency distributions). The first thing that needs to be done is to set up a data file that contains the variables (and their codes) listed in the equation.

6.3.8 Data Entry Format

The data set has been saved under the name: **EX8.SAV**

Variables	Codes
f	Frequency of each interval
x	Midpoint of interval

6.3.9 SPSS Syntax Method

```
COMPUTE fx = f*x.
COMPUTE MEAN = (308+441+420+1197+1638+1785+2576+1260+784+637+
420)/100.
EXECUTE.
```
(Note that the score values are the computed *fx* values.)

1. From the menu bar, click **File,** then **New,** and then **Syntax.** The following **IBM SPSS Statistics Syntax Editor** Window will open.

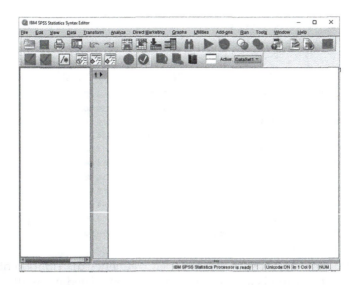

2. Type the **Compute** syntax command in the **IBM SPSS Statistics Syntax Editor** Window.

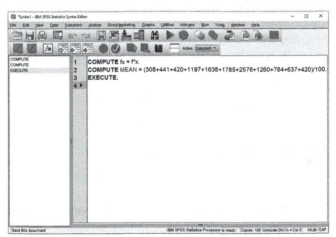

3. To run the **Compute** command, click ▶ or click Run and then **All**.

6.3.10 SPSS Output

Successful execution of the **Compute** command will yield the mean from the grouped frequency distribution of the 100 IQ scores, which is added to the **EX8.SAV** data file under the variable name **MEAN.** The following **Data View** screen shows the mean from the grouped frequency distribution of the 100 IQ scores.

Visible: 4 of 4 Variables

	x	f	fx	MEAN	var	val
1	154.00	2.00	308.00	114.66		
2	147.00	3.00	441.00	114.66		
3	140.00	3.00	420.00	114.66		
4	133.00	9.00	1197.00	114.66		
5	126.00	13.00	1638.00	114.66		
6	119.00	15.00	1785.00	114.66		
7	112.00	23.00	2576.00	114.66		
8	105.00	12.00	1260.00	114.66		
9	98.00	8.00	784.00	114.66		
10	91.00	7.00	637.00	114.66		
11	84.00	5.00	420.00	114.66		

The computed **MEAN** from the grouped frequency distribution of the 100 IQ scores is **114.66** and is identical to the value calculated manually.

6.3.11 The Overall Mean

There will be occasions when we may want to calculate the overall mean from several groups of data. In such a case, we simply sum all the scores from the data sets and divide the summed scores by the total number of scores. That is,

$$\bar{x}\ \text{overall} = \frac{\text{sum of all scores}}{N}$$

$$\bar{x}\ \text{overall} = \frac{\sum X_1(\text{first group}) + \sum X_2(\text{second group}) + \cdots + \sum X_i(\text{last group})}{n_1 + n_2 + \cdots n_i}$$

This can be easily done if the data sets are small. For example, consider the three data sets below (each containing 3 scores).

A	B	C
2	8	14
4	10	16
6	12	18
$\Sigma A = 12$	$\Sigma B = 30$	$\Sigma C = 48$
$\bar{x} = 4$	$\bar{x} = 10$	$\bar{x} = 16$

To find the overall mean for these three data sets enter the summed values $(\Sigma A, \Sigma B, \Sigma C)$ into the equation. Thus,

$$\bar{x}\,\text{overall} = \frac{\Sigma A + \Sigma B + \Sigma C}{nA + nB + nC}$$

$$\bar{x}\,\text{overall} = \frac{12 + 30 + 48}{3 + 3 + 3}$$

$$\bar{x}\,\text{overall} = \frac{90}{9} = 10$$

Therefore, the overall mean for the three data sets is 10. As mentioned earlier, this equation works well for small data sets. However, if the data sets contain many scores then summing the scores for each data set can be cumbersome. For example, if data set A contains 1000 scores, data set B contains 1500 scores, and data set C contains 2000 scores, summing the scores in each data set can be time consuming. There is however a shortcut if we know the mean of each data set and the number of scores in each data set. This shortcut can be demonstrated using the three data sets above (A, B, C) each containing 3 scores.

Since $\bar{x} = (\Sigma X)/n$ we can calculate ΣX by multiplying $\bar{x}$ by n, that is, $n(\bar{x})$. Thus, substituting ΣX with $n(\bar{x})$ in the above 'overall mean' equation we have

$$\bar{x}\,\text{overall} = \frac{\Sigma X_1 + \Sigma X_2 + \Sigma X_3}{n + n_2 + n_3}$$

$$\bar{x}\,\text{overall} = \frac{n_1(\bar{x}_1) + n_2(\bar{x}_2) + n_3(\bar{x}_3)}{n_1 + n_2 + n_3}$$

$$\bar{x}\,\text{overall} = \frac{3(4) + 3(10) + 3(16)}{3 + 3 + 3}$$

$$\bar{x}\,\text{overall} = \frac{12 + 30 + 48}{3 + 3 + 3}$$

$$\bar{x}\,\text{overall} = \frac{90}{9} = 10$$

This shortcut has produced the overall mean of 10 for the three data sets combined, which is identical to the overall mean value calculated from the original equation.

6.3.12 An Example

Let's try another example. Suppose an investor bought three lots of OILCOM shares at different share prices. The price per share and the number of shares bought across the three lots are as follows:

	OILCOM		
	Lot 1	Lot 2	Lot 3
Price per share	$120	$95	$115
N of shares bought	1500	3000	1000
Mean prices ($\bar{x}$) for the three lots of shares	$120	$95	$115

What is the overall mean price for the three lots of shares combined? To solve this problem apply the shortcut equation above.

$$\bar{x}\,overall = \frac{n_1(\bar{x}_1)+n_2(\bar{x}_2)+n_3(\bar{x}_3)}{n_1+n_2+n_3}$$

$$\bar{x}\,overall = \frac{1500(120)+3000(95)+1000(115)}{1500+3000+1000}$$

$$\bar{x}\,overall = \frac{180,000+285,000+115,000}{1500+3000+1000}$$

$$\bar{x}\,overall = \frac{580,000}{5500} = 105.45$$

Therefore, the overall mean price for the three lots of OILCOM shares combined is $105.45 per share.

6.3.13 How to Calculate the Overall Mean Using SPSS

We can calculate the overall mean price for the example above using the SPSS syntax method (SPSS does not provide the Windows method for calculating overall mean values). The first thing that needs to be done is to set up a data file that contains the variables and their values listed in the equation.

6.3.14 Data Entry Format

The data set has been saved under the name: **EX9.SAV**

Variables	Value
Lot 1	120
Lot 2	95
Lot 3	115
N1	1500
N2	3000
N3	1000

6.3.15 SPSS Syntax Method

```
COMPUTE OVERALL_MEAN = ((N1*LOT1) + (N2*LOT2) + (N3*LOT3))/
(N1 + N2 + N3).
EXECUTE.
```

1. From the menu bar, click **File**, then **New**, and then **Syntax.** The following **IBM SPSS Statistics Syntax Editor** Window will open.

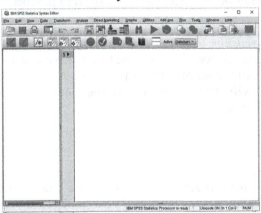

2. Type the **Compute** syntax command in the **IBM SPSS Statistics Syntax Editor** Window.

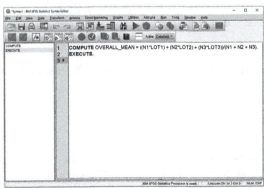

3. To run the **Compute** command, click ▶ or click Run and then **All.**

6.3.16 SPSS Output

Successful execution of the Compute command will yield the overall mean price for the three lots of shares combined, which is added to the EX9.SAV data file under the variable name **OVERALL_MEAN**. The following Data View screen shows the overall mean price for the three lots of shares combined is **$105.45**, which is identical to the value calculated manually (see Section 6.3.12).

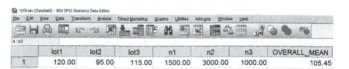

6.3.17 Properties of the Mean

There are properties of the mean that are useful in practice. These properties are as follows:

1. *The mean is sensitive to the exact values of all the scores in the distribution.* This property derives from the fact that the mean is calculated on the basis of all the scores in a distribution. As such, a change in one score value will change the value of the mean. Consider the following set of scores:

$$2,4,6,8,10 \quad \bar{x} = \frac{\sum X}{N} = \frac{30}{5} = 6.00$$

   The mean for this set of scores is 6. If we change the value of one score, the value of the mean will change. Let's say we change the number 2 to number 4 in the above set of scores. This will change the mean value to 6.4.

$$4,4,6,8,10 \quad \bar{x} = \frac{\sum X}{N} = \frac{32}{5} = 6.40$$

2. *The sum of the deviations of all the values of X from their arithmetic mean is zero.*

$$\sum(X - \bar{x}) = 0$$

   This property states that if the mean value ($\bar{x}$) of a distribution of scores is subtracted from each score, the sum of the differences will equal zero. This property can be demonstrated using the scores presented in Table 6.3.

TABLE 6.3

Demonstration of the Property that $\Sigma(X-\bar{x})=0$

X	$(X-\bar{x})$	$\bar{x}$
1	−4	$5\left(\bar{x}=\dfrac{\Sigma X}{N}=\dfrac{25}{5}=5.00\right)$
3	−2	
5	0	
7	+2	
9	+4	
$\Sigma X=25$	$\Sigma(X-\bar{x})=0$	

This property implies that the mean is the "balance point" of the set of scores. Imagine, for example, you are sitting at the mean, with the negative numbers to your left and the positive values to your right. If the scores are in balance, then the sum of the negative numbers (−6) must equal the sum of the positive numbers (+6) and the difference is therefore 0. If they were unequal, the difference between the sum of the negative numbers and the sum of the positive numbers would not be zero. Since the difference is zero, the positive and negative scores must be equal or in balance.

3. *The mean is very sensitive to extreme scores when these scores are not balanced on both sides of it.* This property derives from the fact that the mean is the balance point of the distribution. Consider the following two sets of scores:

$$2,4,6,8,10 \quad \bar{x}=\frac{\Sigma X}{N}=\frac{30}{5}=6.00$$

$$2,4,6,8,100 \quad \bar{x}=\frac{\Sigma X}{N}=\frac{120}{5}=24.00$$

It can be seen that all the scores in the two distributions are the same except for the very large score of 100 in the second distribution. This one extreme score has not only changed the mean but also has quadrupled it!

4. *The sum of the squared deviations from the mean is less than the sum of the squared deviations about any other scores. That is, the sum of the squared deviations from the mean is a minimum.*

$$\Sigma(X-\bar{x})^2 \quad \text{is a minimum}$$

This property implies that the mean is that measure of central tendency, which makes the sum of the squared deviations around it a

TABLE 6.4

Demonstration of the Property that $\Sigma(X-\bar{x})^2$ is a Minimum

X	$(X-\bar{x})^2$	$(X-2)^2$	$(X-3)^2$	$(X-4)^2$
1	16	1	4	9
3	4	1	0	1
5	0	9	4	1
7	4	25	16	9
9	16	49	36	25
$\Sigma X=25$	$\Sigma(X-\bar{x})^2=40$	$\Sigma(X-2)^2=85$	$\Sigma(X-3)^2=60$	$\Sigma(X-3)^2=45$
$N=5$				
$\bar{x}=5$				

minimal. This is an important characteristic of the mean as it is used in many areas of statistics, particularly in regression analysis. This property of the mean can be demonstrated in Table 6.4, which shows a set of raw scores (X), the squared deviations of the scores from the mean $(X-\bar{x})^2$, and the sum of the squared deviations $(\Sigma(X-\bar{x})^2)$ when the deviations are taken from the mean and from various other scores.

It can be seen that the sum of the squared deviations is smallest when the deviations are taken from the mean.

5. *Under ordinary circumstances, the mean is the most resistant to chance sampling variation.* If we draw a series of samples at random from a large population of scores, the means of these samples would be different (because of the effects of chance). This would also be true for the medians and the modes. However, the mean varies less than these other measures of central tendency. Consider the following two sets of scores drawn from a large population:

$$9, 15, 24, 26, 26 \quad \Sigma X=100$$
$$N=5$$
$$\bar{x}=20$$
$$\text{Median}=24$$
$$\text{Mode}=26$$

$$5, 15, 20, 30, 30 \quad \Sigma X=100$$
$$N=5$$
$$\bar{x}=20$$
$$\text{Median}=20$$
$$\text{Mode}=30$$

It can be seen from the two sets of the scores that their mean values are the same ($\overline{x} = 20$). However, the median and mode values are different. This implies that the mean, as a measure of central tendency, is more stable from sample to sample than either the median or the mode. This is an important property of the mean because sampling stability is a major requirement in inferential statistics and is a major reason why the mean (rather than the median or the mode) is used in inferential statistics whenever possible.

6.4 The Median

The median of a set of scores is the middle number in the set. It is the number that is halfway into the set and therefore divides the set of scores into two halves.

6.4.1 Calculating the Median for Ungrouped Scores

When dealing with ungrouped scores, it is quite easy to find the median. The scores are first rank-ordered from smallest to largest. The median is the "middle" value in the list of numbers. Calculation of the median for ungrouped scores actually depends on whether the data set contains an odd number of scores or an even number of scores. When the data set contains an odd number of scores, the median is the middle value in the list of numbers. Consider the data set below containing an odd number of scores ($N = 5$).

X	Rank-Order Median = 5
3	1
7	3
5	5
9	7
1	9

As the median is the middle value in the rank-ordered list of numbers, the median value is 5.

Calculation of the median for an even number of scores is somewhat different. In this case, we find the middle pair of numbers and then find their mean value. This is easily done by adding the middle pair of numbers and dividing by two.

$$\overline{x} = \frac{\Sigma \text{middle pair of numbers}}{2}$$

Consider the data set below containing an even number of scores ($N = 6$).

X	Rank-Order Median = 6
3	1
7	3
5	5
9	7
1	9
11	11

To find the median add the two middle numbers and divide by 2, that is, ($5 + 7/2$). The median value for this even set of scores is therefore **6**.

If the median occurs at a value where there are tied scores, the tied score will be used as the median. Consider the data set below containing an odd number of scores ($N = 7$) with 4 as the tied score.

X	Rank-Order Median = 4
3	2
7	3
5	4
9	4
4	5
4	7
2	9

As the tied score will be used as the median, the median for this set of scores is **4**.

6.4.2 Calculating the Median for Grouped Scores

With grouped scores, the median is defined as *that score below which 50% of the scores in the distribution fall*. If this definition sounds familiar, it is because the median for grouped scores is simply the *50th percentile* (P_{50}). To calculate the median (P_{50}), follow the steps for the manual calculation and the SPSS syntax method described in Chapter 4.

6.4.3 Properties of the Median

There are two properties of the median that are noteworthy.

First, *the median is insensitive to extreme scores*.

Unlike the mean, the median is not affected by extreme scores. To demonstrate this characteristic, consider the three sets of scores presented in Table 6.5.

TABLE 6.5

Demonstration of the Effect of Extreme
Scores on the Mean and Median

Scores	Mean	Median
2, 4, 6, 8, 10	6	6
2, 4, 6, 8, 100	24	6
2, 4, 6, 8, 1000	204	6

The three sets of scores are similar except for the extreme scores in the second set of scores (100) and the third set of scores (1000). As the mean is sensitive to extreme scores (it is calculated on the basis of every score in the distribution), the mean values of the three sets of scores have changed significantly. This is not the case for the median. As the median is calculated as a single middle number, its value is not affected by the extreme scores. For this reason, when the distribution is affected by extreme scores resulting in a highly *skewed* distribution, it is better to represent the central tendency with the median than the mean. The following example demonstrates this point quite easily.

Say you are interested in buying a home in a particular neighborhood and you ask the real estate agent what the average price of the houses is. Let's suppose that there are 10 houses in this neighborhood and their prices are as follows:

House	Mean	Median
$125,000	$330,700	$182,500
$130,000		
$150,000		
$160,000		
$180,000		
$185,000		
$190,000		
$192,000		
$195,000		
$1,800,000		

You can see that the prices of 9 of the 10 houses range from $125,000 to $195,000 with one house priced at $1,800,000. The extreme price of this house has skewed the distribution of prices and produced a mean price of $330,700. This mean price is clearly not representative of the average price of 90% of the houses in the neighborhood. The median price of $182,500 (which is not affected by the extreme price) is more representative of the house prices in the neighborhood. This is why for highly skewed distributions, the median is better at representing the average than the mean.

Second, *under normal circumstances, the median is subject to greater sampling variation than the mean. That is, it is not as stable as the mean from sample to sample.*

For example, if we draw numerous samples from the same population, the means and the medians of these different sets of scores will change. However, the median is subject to greater variability than the mean. This follows from the simple fact that the median is calculated as a single middle number, whereas the mean is calculated from all the scores in the set of scores. Consider the three sets of scores below.

Scores	Mean	Median
2, 4, 6, 8, 10	6	6
1, 2, 3, 9, 15	6	3
2, 4, 5, 9, 10	6	5

While the three sets of scores are different, the means are the same, but the medians are different. This shows that when sampling from a population, the median is subject to greater sampling variability (less stable) than the mean, which renders it less useful in *inferential* statistics.

6.5 The Mode

In a frequency distribution, the mode is the value that occurs most often. If no number is repeated, then there is no mode for the list. Of the measures of central tendency, it is obvious that the mode is the most easily determined as it is obtained by "eye-balling" the frequency distribution rather than by computation. For example, consider the set of scores below.

Scores: 3, 7, 5, 9, 4, 4, 2

Rank-ordered: 2, 3, 4, 4, 5, 7, 9

Rank-ordering the scores makes them easy to see which numbers appear most often. In this case, the mode is 4.

Some distributions can have more than one mode. For example, consider the set of scores below.

2, 4, 4, 4, 5, 5, 7, 7, 7, 10

Inspection of this set of scores shows that the scores of 4 and 7 each appear three times. As such there are two modes: at **4** and **7**. Having two modes is called "**bimodal.**" Having more than two modes is called "**multimodal.**" The mode can also be easily obtained via graphical form, say from a histogram.

6.5.1 SPSS Windows Method

1. Open the data set **EX10.SAV.** Click **Graphs** on the menu bar, then **Legacy Dialogs**, and then **Histogram**. The following **Histogram** Window will open.

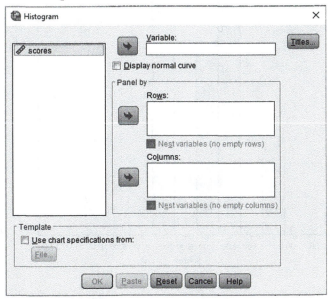

2. Transfer the **SCORES** variable to the **Variable:** cell by clicking the **SCORES** variable (highlight) and then 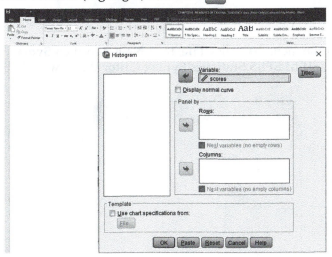.

Click [OK] to draw a histogram of the frequency distribution representing the **SCORES** variable (see Figure 6.1).

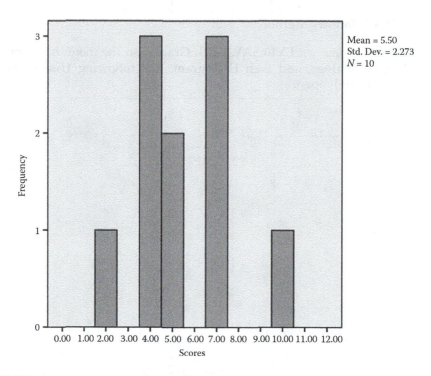

Mean = 5.50
Std. Dev. = 2.273
N = 10

FIGURE 6.1
Histogram of the frequency distribution representing the SCORES variable.

6.5.2 SPSS Syntax Method

```
GRAPH
/HISTOGRAM=SCORES.
```

1. From the menu bar, click **File**, then **New**, and then **Syntax.** The following **IBM SPSS Statistics Syntax Editor** Window will open.

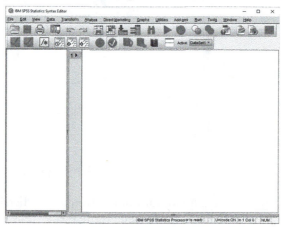

2. Type the **Histogram** syntax command in the **IBM SPSS Statistics Syntax Editor** Window.

3. To run the **Histogram** analysis, click ▶ or click Run and then **All**.

6.5.3 SPSS Histogram Output

Figure 6.1 presents the histogram of the frequency distribution representing the **SCORES** variable. It can be seen that the scores of 4 and 7 have the highest peaks (each appear three times). As such there are two modes: at **4** and **7**.

6.5.4 The Mode for Grouped Scores

For grouped scores, the mode is the midpoint of the interval containing the highest frequency count. In Table 4.5, the mode is the score of 112 since it is the midpoint of the interval (109–115) containing the highest frequency count ($n = 23$). The lower case n is the number of a subsample whereas N is the total number of the sample.

6.6 Comparison of the Mean, Median, and Mode

In general, the arithmetic mean is the preferred statistic for representing central tendency. This preference is based on a number of properties it possesses. First, the concept of *'deviation from the mean'* gives rise to two of its most important properties, that is, *the sum of the deviation scores from the mean is zero*, and *the sum of the squared deviation scores from the mean is a minimum*. Thus, deviations of scores from the mean provide valuable information about any distribution. In contrast, deviation scores from the median and the corresponding squared deviations from the median have limited statistical applications, both practically and theoretically.

Second, as mentioned earlier, the arithmetic mean is more stable as a measure of central tendency than either the median or the mode. If samples were repeatedly drawn from a given population, the means of these samples will vary, but they will vary less than either the medians or the modes. In other words, the mean is a more stable (and therefore a more reliable) measure of its population central tendency.

On the other hand, there are situations where the median is the preferred measure of central tendency over the mean. In general, when the distribution of scores is symmetrical (appearing as a normal bell-shaped curve), the mean and median are identical and the mean should be used as the preferred measure of central tendency. However, when the distribution of scores is highly skewed (see Table 6.5), the mean will provide a misleading estimate of the central tendency. From Table 6.5, the first set of scores is balanced in that it is not affected by extreme scores. This balanced set of scores produced the identical mean and median values of 6. However, the second and third sets of scores are affected by high extreme scores resulting in the fact that their means no longer reflect accurately the majority of the scores in the distributions, that is, these means have overestimated the majority of the scores in the distributions. Under these circumstances, the median is the preferred measure of central tendency.

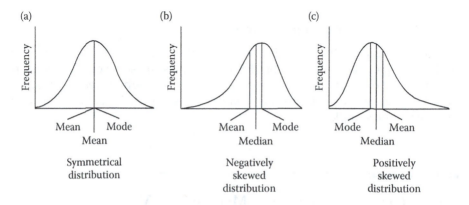

FIGURE 6.2
Relationship between the mean, median, and mode in (a) symmetrical, (b) negatively, and (c) positively skewed distributions.

The mode is most often used when the researcher requires a quick and rough estimate of the distribution's central tendency. It is rarely used in the behavioral sciences.

6.7 Measures of Central Tendency: Symmetry and Skewness

In a unimodal symmetrical frequency distribution, the mean, median, and mode will be equal. This is shown in the bell-shaped curve in Figure 6.2a. When the distribution is skewed, the mean and median will not be equal. The basic fact to keep in mind is that the mean is most affected by extreme scores and therefore is drawn in the direction of the skew; the median, unaffected by extreme scores is not. Thus, when the mean is higher than the median (as in the case of extreme high scores), the distribution can be said to be *positively skewed* (Figure 6.2c); when the mean is lower than the median (as in the case of extreme low scores), the distribution can be said to be *negatively skewed* (Figure 6.2b). Figure 6.2 shows these relationships.

7

Measures of Variability/Dispersion

7.1 What Is Variability?

The terms *variability*, *spread*, and *dispersion* are synonymous, and refer to how spread out a distribution of scores is. Although measures of central tendency focus on how scores in a distribution are congregated around the middle of the distribution, measures of variability are concerned with how the scores are spread out or dispersed along the distribution. Why is variability important? In the social sciences, variability serves two major goals. First, many of the statistical inferential tests employed for testing hypotheses require knowledge of the variability of the scores. For example, in an investigation on gender difference in IQ scores, the researcher may ask, "Is there a difference between males and females in their IQ scores?" The two groups' distributions of IQ score can be represented by the two "bell-shaped" curves presented in Figure 7.1.

The mean IQ scores for both groups are indicated with dashed lines. Looking at the two different possible outcomes labeled *high* and *low* variability, we can see that the difference between the means in the two situations is exactly the same. The only thing that differs between these is the variability or "spread" of the scores around the means, that is, low versus high variability. Of these two cases, which one would make it easier for us to conclude that the mean IQ scores of the two groups are different? The answer is the low variability case. This is because the lower the variability (i.e., the more consistently the scores are congregated around the groups' respective means) the less is the amount of overlap between the bell-shaped curves for the two groups. Alternatively, in the high variability case, it can be seen that there is a high amount of overlap between the two curves denoting the fact that there are quite a few males whose IQ scores are in the range of the females and vice versa. So why is this important? Because if we are interested in investigating group differences, it is not sufficient to simply consider the mean differences between the groups – we need to also consider the variability around the groups' means. Thus, even if there is very little difference between the means of two groups, if the variability of the distributions of scores is low (low amount of overlap between the curves), then there is a high

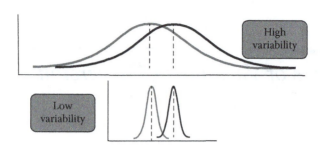

FIGURE 7.1
Demonstration of low and high variability.

probability that the two groups will differ. On the other hand, a large difference between the means of two groups may not denote 'group difference' if the variability of the distributions of scores is high (high amount of overlap between the curves).

In other words, a large difference between means will be hard to detect if there is a lot of variability in the scores' distributions.

Second, understanding variability is important because it is useful in its own right. For example, say you were one of the 100 participants who took an IQ test (see Table 4.2) and you scored an IQ score of 125. As mentioned earlier, a score by itself is meaningless and takes on meaning only when it is compared with other scores or other statistics. The measure of central tendency provides a standard against which to compare your IQ score of 125. For example, if you know the mean of the IQ scores, you can determine whether your score is higher or lower than the mean score. But how much higher or lower? As useful as the mean is as a measure of central tendency, it does not provide an answer to these questions. To fully interpret your IQ score of 125, it is clear that additional information concerning the variability or dispersion of the entire set of IQ scores about the mean is required. What is the average deviation/variability of the scores about the mean? What is the extent of the spread of the set of IQ scores as determined by the lowest and highest scores? Given the mean and average deviation scores, what is the *percentile rank* of your IQ score of 125? The mean, working in conjunction with the measures of variability, will provide answers to these questions.

Three measures of variability are commonly used in the social sciences: the *range*, the *standard deviation*, and the *variance*.

7.2 Range

The range is the easiest measure of variability to calculate, and is defined as the difference between the highest score and the lowest score in a distribution. Therefore,

$$\text{Range} = \text{highest score} - \text{lowest score}$$

Consider the set of 100 IQ scores presented in Table 4.2. The highest score is 155 and the lowest number is 81, so $155 - 81 = 74$. The range (spread of scores) is 74 units wide. Let's take another example. Consider the following 10 numbers: 109, 55, 33, 77, 55, 101, 92, 88, 72, 61. What is the range? The highest number is 109 and the lowest number is 33, so $109 - 33 = 76$; the range is 76.

7.3 Standard Deviation

The standard deviation (σ), or sigma, of a set of scores is defined as the *average deviation of the set of scores about the mean of the distribution*. As the standard deviation is concerned with the average deviation score, it is best to start off our discussion of the standard deviation by discussing what deviation scores are. A deviation score is simply the difference between a raw score and the mean of its distribution, that is, $(X - \bar{x})$. It tells us how many units the raw score is above or below the mean. For example, consider the set of raw scores (X) presented in Table 7.1.

Once the raw scores (X) have been transformed into deviation scores $(X - \bar{x})$, the transformation tells us many units a raw score is above or below the mean. Thus, the score of 1 lies 4 units below the mean, and the score of 9 lies 4 units above the mean.

7.3.1 Calculating the Standard Deviation Using the Deviation Scores Method

As mentioned earlier, the standard deviation is simply the average deviation score about the mean of a set of scores. By this definition, the calculation of the standard (average) deviation follows the same logic as calculating the

TABLE 7.1

Deviation Scores

X	$X - \bar{x}$	$\bar{x}$
1	$1 - 5 = -4$	$5\left(\bar{x} = \dfrac{\Sigma X}{N} = \dfrac{25}{5} = 5.00\right)$
3	$3 - 5 = -2$	
5	$5 - 5 = 0$	
7	$7 - 5 = +2$	
9	$9 - 5 = +4$	

average or mean of a set of raw scores. Thus, if the equation for calculating the mean of raw scores is

$$\bar{x} = \frac{\sum X}{N}$$

it follows that the equation for calculating the mean of the deviation scores (the standard deviation) is

$$\bar{x} \text{ deviation score} = \frac{\sum(X - \bar{x})}{N}$$

This logic is correct except for a major stumbling block. Recall that an important property of the mean is that *the sum of the deviations of all the values of X from their arithmetic mean is zero*. That is,

$$\sum(X - \bar{x}) = 0$$

The proof of this property is demonstrated by summing all the deviation scores in Table 7.1, which is zero. Thus, regardless of the magnitude of the deviation scores in any distribution, the sum of the deviation scores will always equal zero. This property of the mean makes it impossible to calculate the standard deviation from the equation

$$\bar{x} \text{ deviation score} = \frac{\sum(X - \bar{x})}{N}$$

as it will produce a standard deviation of zero for any data set. This is because *the negative deviation scores are cancelling out the positive deviation scores.*

The solution to this problem is to transform the *negative* deviation scores into *positive* deviation scores by squaring the negative deviation scores. Note that squaring transforms negative numbers into positive numbers. Thus, after squaring there will only be positive deviation scores and their sum will no longer be equal to zero! Table 7.2 presents the squared deviation scores and the sum of these scores.

As can be seen from Table 7.2, the sum of the squared deviation scores $[\sum(X - \bar{x})^2]$ no longer equals zero – the summed value is 40. To obtain the average value, we simply divide the sum of the squared deviation scores (40) by N, that is, $\sum(X - \bar{x})^2 / N$. Once again, this logic is correct but unfortunately this equation has produced the *mean squared deviation score* and not the *mean deviation score* (the standard deviation). Since this problem is caused by the fact that the equation required the squaring of the deviation scores, the solution to this problem involves simply 'unsquaring' the

TABLE 7.2

Squared Deviation Scores

X	$X - \bar{x}$	$(X - \bar{x})^2$	$\bar{x}$
1	$1 - 5 = -4$	16	$5\left(\bar{x} = \dfrac{\sum X}{N} = \dfrac{25}{5} = 5.00\right)$
3	$3 - 5 = -2$	4	
5	$5 - 5 = 0$	0	
7	$7 - 5 = +2$	4	
9	$9 - 5 = +4$	16	
		$\sum(X - \bar{x})^2 = 40$	

equation. This is done by taking the square root of $\sum(X-\bar{x})^2/N$ (mean squared deviation score.)

$$S_x = \sqrt{\frac{\sum(X - \bar{x})^2}{N}}$$

And there we have it – the equation for calculating the standard deviation! However, it should be noted that while this equation is technically correct, it is the equation for calculating the standard deviation for a set of *population* scores. In research though, we tend to use sample scores to calculate the standard deviation with the aim of using this calculation to estimate the standard deviation of the population. The present equation with N as the denominator (when applied to sample scores) tends to yield a standard deviation value that is too small and therefore underestimates the population standard deviation. To compensate for this underestimation, we make the denominator smaller by subtracting 1 from N, that is, $N - 1$. Thus, the equation for calculating the standard deviation for a set of sample scores is

$$S_x = \sqrt{\frac{\sum(X - \bar{x})^2}{N - 1}}$$

Table 7.3 demonstrates the calculation of the standard deviation for the sample data set presented in Table 7.1

$$S_x = \sqrt{\frac{\sum(X - \bar{x})^2}{N - 1}}$$

$$S_x = \sqrt{\frac{40}{4}}$$

TABLE 7.3

Calculation of the Standard Deviation for the Sample Data Set Presented in Table 7.1

X	$X - \bar{x}$	$(X - \bar{x})^2$	$\bar{X}$
1	$1 - 5 = -4$	16	$5\left(\bar{x} = \dfrac{\sum X}{N} = \dfrac{25}{5} = 5.00\right)$
3	$3 - 5 = -2$	4	
5	$5 - 5 = 0$	0	
7	$7 - 5 = +2$	4	
9	$9 - 5 = +4$	16	
	$\sum(X - \bar{x})^2 = 40$		

$$S_x = \sqrt{10}$$

$$S_x = 3.16$$

Thus, the standard deviation for the sample data set is 3.16. That is, on average, the sample scores vary about the mean ($\bar{x} = 5.00$) by ±3.16 points.

7.3.2 Calculating the Standard Deviation Using the Raw Scores Method

While the above equation for calculating the standard deviation for a set of sample scores is correct, the actual calculation procedure can be cumbersome if the data set (1) contains many numbers, (2) requires the calculation of the deviation score for each number, and (3) when the deviation scores have decimal remainders. There is a simpler way to calculate the standard deviation that bypasses the need to calculate the deviation scores and the squared deviation scores. This computation method employs the raw scores from the data set.

It can be shown mathematically that

$$\sum(X - \bar{x})^2 = \sum X^2 - \frac{(\sum X)^2}{N}$$

This equation allows for the calculation of the summed squared deviation scores using the data set's raw scores (X). Table 7.4 demonstrates the calculation of the standard deviation using the raw scores method.

The computed standard deviation using the raw scores method is 3.16, which is identical to the value calculated using the deviation scores method (see Table 7.3).

TABLE 7.4

Demonstration of the Calculation of the Standard Deviation Using the Raw Scores Method

X	X^2	$\Sigma(X-\bar{x})^2$	S_x
1	1	$=\Sigma X^2 - \dfrac{(\Sigma X)^2}{N}$	$=\sqrt{\dfrac{\Sigma(X-\bar{x})^2}{N-1}}$
3	9		
5	25	$=165 - \dfrac{(25)^2}{5}$	$=\sqrt{\dfrac{40}{4}}$
7	49		
9	81	$=165-125$	$=\sqrt{10}$
$\Sigma X = 25$	$\Sigma X^2 = 165$		
$N = 5$		$= 40$	$= 3.16$

7.4 Variance

The variance is defined as *the average sum of the squared deviation scores from the mean*. In equation form, the variance is represented by

$$s^2 = \frac{\Sigma(X-\bar{x})^2}{N-1}$$

If this equation looks familiar to you, it is because it is the equation for the standard deviation minus the square root. Thus, the variance is essentially the square of the standard deviation. Because the variance yields squared units of measurement it is used more often in inferential statistics than in descriptive statistics.

7.5 Using SPSS to Calculate the Range, the Standard Deviation, and the Variance

Let's say we want to calculate the range, the standard deviation, and the variance for the set of 100 IQ scores presented in Table 4.2.

7.5.1 SPSS Windows Method

1. Launch the SPSS program and then open the data file **EX1.SAV**. Click **Analyze** on the menu bar, then **Descriptive Statistics**, and then **Frequencies.** The following **Frequencies** Window will open.

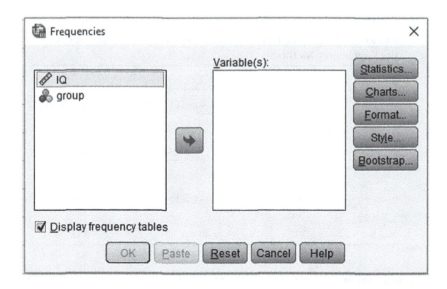

2. In the left-hand field containing the study's **IQ** variable, click (high-light) this variable, and then click 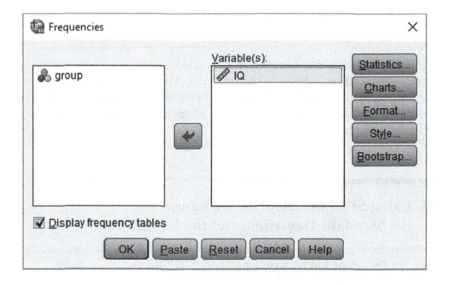 to transfer the selected IQ variable to the **Variable(s)**: field.

3. Click [statistics...] to open the **Frequencies Statistics** Window below.

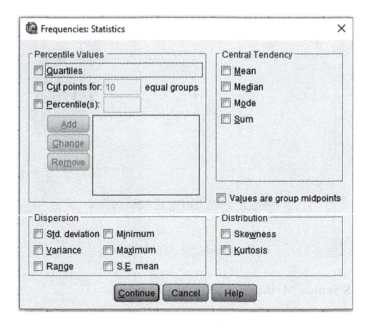

4. Under the **Dispersion** heading, check the **St̲d. deviation** cell, the **V̲ariance** cell, and the **Ra̲nge** cell.

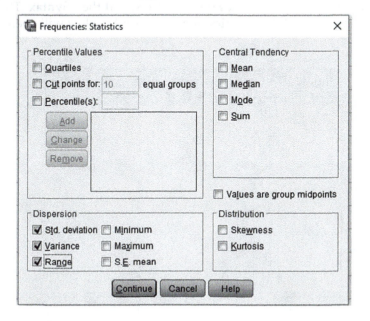

Click Continue to return to the **Frequencies** Window.

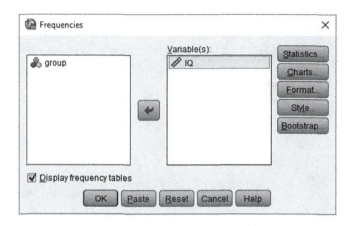

5. Click [OK] to run the analysis. See Table 7.5 for the results.

7.5.2 SPSS Syntax Method

```
FREQUENCIES VARIABLES=IQ
/STATISTICS=STDDEV VARIANCE RANGE
/ORDER=ANALYSIS.
```

1. From the menu bar, click **File**, then **New**, and then **Syntax.** The following **IBM SPSS Statistics Syntax Editor** Window will open.

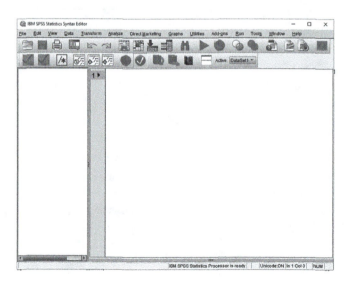

2. Type the **Frequencies** analysis syntax command in the **IBM SPSS Statistics Syntax Editor** Window.

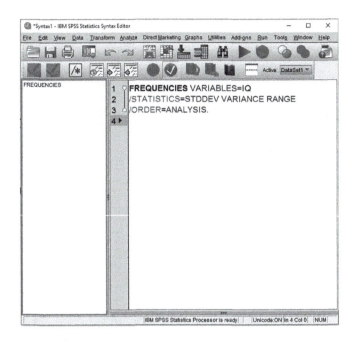

3. To run the **Frequencies** analysis, click ▶ or click Run and then **All**.

7.5.3 SPSS Output

As can be seen from Table 7.5, the standard deviation, the variance, and the range for the set of 100 IQ scores are 16.01, 256.19, and 74, respectively. That is, (1) on average the sample IQ scores vary about the mean ($\bar{x} = 114.54$) by ±16.01 points (standard deviation), (2) the sum of the squared deviation scores is 256.19 (variance), and (3) the spread of the scores from the minimum to the maximum is 74 units wide (range).

TABLE 7.5

Frequencies Output of the Standard Deviation, the Variance, and the Range of the 100 IQ Scores

Statistics		
IQ		
N	Valid	100
	Missing	0
Std. deviation		16.00595
Variance		256.190
Range		74.00

8

The Normal Distribution and Standard Scores

8.1 The Normal Distribution

The normal distribution, sometimes called the *normal curve* or *bell-shaped curve*, is a graphical display of a set of scores whose distribution approximates a bell curve. It is a distribution that occurs naturally in many situations, such as people's height, weight, IQ scores, academic grades, and salaries. Take height as an example. The majority of people in a population will be of average height, say 172 cm. There will be very few people who are giants (much taller than 172 cm) and very few people who are very short (much less than 172 cm). This creates a symmetrical distribution that resembles a bell in which half of the data will fall to the left of the mean and half to the right. Moreover, in a symmetrical bell curve distribution, the mean, median, and mode are all equal.

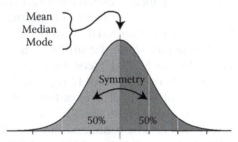

Many physical characteristics follow this type of pattern and the resultant bell curve is the main reason why it is widely used in statistics and in many social/behavioral sciences.

8.2 Areas Contained under the Standard Normal Distribution

It is possible to calculate any area contained under the standard normal (bell-shaped) distribution from its mean (μ) and standard deviation (σ).

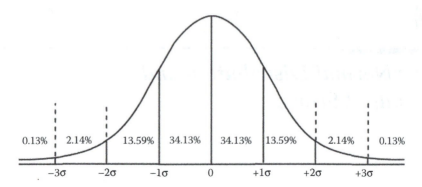

FIGURE 8.1
Areas under the normal distribution in relation to the distribution's μ and σ.

The normal distribution has a μ of 0, a σ of 1, and a total area equal to 1.00. The relationship between the μ and σ with regard to the area under the normal curve is depicted in Figure 8.1.

From Figure 8.1, it can be seen that 34.13% of the area under the curve is contained between the mean and 1 standard deviation above the mean (μ + 1σ). Similarly, 34.13% of the area under the curve is contained between the mean and 1 standard deviation below the mean (μ − 1σ). It can be seen that 47.72% (34.13% + 13.59%) of the area under the curve is contained between the mean and 2 standard deviations above the mean (μ + 2σ); a similar 47.72% of the area under the curve is contained between the mean and 2 standard deviations below the mean (μ − 2σ). It can also be seen that 49.86% (34.13% + 13.59% + 2.14%) of the area under the curve is contained between the mean and 3 standard deviations above the mean (μ + 3σ); a similar 49.86% of the area under the curve is contained between the mean and 3 standard deviations below the mean (μ − 3σ).

Let's use an example to demonstrate the utility of the above information. Suppose we have a population of 20,000 IQ scores that are normally distributed with a μ of 100 and a σ of 12 (note that a σ of 12 represents 1 σ above or below the μ). Thus, 34.13% or 6,826 (0.3413 × 20,000) of the scores will be contained between the scores of 100 (μ) and 112 (μ + 1σ); 47.72% (34.13% + 13.59%) or 9,544 (0.4772 × 20,000) of the scores will be contained between the scores of 100 (μ) and 124 (μ + 2σ); 49.86% (34.13% + 13.59% + 2.14%) or 9,972 (0.4986 × 20,000) of the scores will be contained between the scores of 100 (μ) and 136 (μ + 3σ).

8.3 Standard Scores (z Scores) and the Normal Curve

Let's suppose that you are one of 50 students who took a statistics test and your score is 85 out of 100. A score of 85 is simply a number and is virtually meaningless unless you have a standard against which to compare your score. Without such a standard, you will not be able to tell whether your score of 85 is high, moderate, or low. However, if you know the mean and standard deviation for the set of 50 statistics test scores, you will be able to calculate, for example, the *percentile rank* of your test score of 85, that is, the percentage of scores that is lower than your score of 85. In other words, you will be able to calculate how well you performed compared to the other 50 students in your statistics class.

Before tackling the above problem let us look at what standard scores or z scores are and how they can be used to solve the above problem. When a frequency distribution is normally distributed, we can calculate the area under the normal curve between a particular score and the mean of its distribution by standardising the scores, that is, by converting the distribution's raw scores to *standard scores* (or z scores).

> *A standard score or z score depicts the number of standard deviations from (above or below) the distribution's mean.*

In other words, a standard or z score indicates how many standard deviations a score is above or below the mean of the distribution. It should be noted that when raw scores are converted into z scores, the z scores' distribution has a mean of 0 and a standard deviation of 1.

Simply knowing a z score offers no information about its corresponding raw score. Its utility lies in comparative studies where a z score can indicate how well a person did compared to other test-takers in the same or norm or reference group. This is particularly beneficial to educators interested in knowing how a particular student performed relative to other students in the same or different class. Z scores range from −3 standard deviations to +3 standard deviations from the mean, with the mean equalling 0. Therefore, positive z scores indicate the number of standard deviations above the mean while negative z scores indicate the number of standard deviations below the mean (see Figure 8.1).

The formula for calculating a z score is relatively straightforward. Keeping in mind that a z score indicates the number of standard deviations a raw score deviates from its distribution mean, the formula for calculating a z score is

$$z \text{ score} = (\text{raw score} - \text{mean}) / \text{standard deviation}$$

or

$$z \text{ score} = \frac{X - \mu}{\sigma}$$

For example, if a distribution of test scores has a mean (μ) of 35 and a standard deviation (σ) of 5, a score of 50 has a z score value of

$$\frac{50 - 35}{5} = 3$$

That is, the raw score of 50 is 3 standard deviations above the mean.

8.3.1 Calculating the Percentile Rank with *z* Scores

Let's go back to our practical example where you want to find out how well you did on a statistics exam. Recall that there are 50 students in your class who took the test and that your test score is 85. You want to calculate the *percentile rank* of your test score of 85, that is, the percentage of scores that is lower than your score of 85. In other words, you want to calculate how well you performed compared to the other 50 students in your statistics class. Let's assume that the set of 50 test scores is normally distributed with a mean of 66.70 and a standard deviation of 10.31 marks.

8.3.2 SPSS Windows Method

1. Launch the SPSS program and then open the data file **EX11.SAV** (this file contains the 50 statistics exam scores). Click **Analyze** on the menu bar, then **Descriptive Statistics**, and then **Descriptives.** The following **Descriptives** Window will open. Check the **Save standardized values as variables** cell.

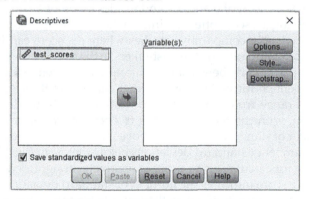

2. In the left-hand field containing the study's **test_scores** variable, click (highlight) this variable, and then click ⏩ to transfer the selected **test_scores** variable to the **Variable(s):** field.

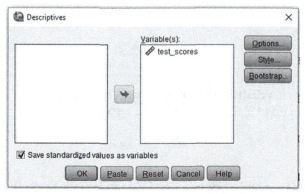

3. Click ⬛ to run the analysis. In running this analysis, SPSS will transform the set of 50 raw scores into z (standard) scores and will append these z scores as a new variable with the name **Ztest_scores** in the data set. Table 8.1 presents the first 10 raw scores and their corresponding computed z scores.

8.3.3 SPSS Syntax Method

```
DESCRIPTIVES VARIABLES=TEST_SCORES
/SAVE
/STATISTICS=MEAN STDDEV MIN MAX.
```

1. From the menu bar, click **File**, then **New**, and then **Syntax**. The following **IBM SPSS Statistics Syntax Editor** Window will open.

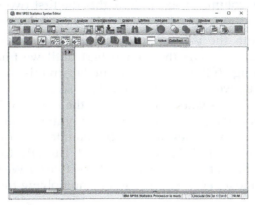

2. Type the **Descriptives** analysis syntax command in the **IBM SPSS Statistics Syntax Editor** Window.

3. To run the **Descriptives** analysis, click ▶ or click **Run** and then **All**.

8.3.4 SPSS Data File Containing the First 10 Computed z Scores

As can be seen from Table 8.1, your statistics test score of **85** has a transformed z score of **1.78** (rounding the number up to two decimal points). That is, your exam score of 85 is 1.78 deviations above the mean. Now, we need to work out the percentage (or number) of students who scored lower than your score (percentile rank). To do this, we need to refer to the z score distribution table (see Table A in the Appendix). This table helps us to identify the percentage of scores that is greater or lesser than your z score. To use the table follow these steps:

1. Recall that your z score value is 1.78 (i.e., 1.78 deviations above the mean).
2. The vertical column (y-axis) highlights the first two digits of the z score and the horizontal bar (x-axis) the second decimal place.
3. We start with the y-axis (vertical column), finding 1.7, and then move along the x-axis (horizontal bar) until we find 0.08, before finally reading off the appropriate number; in this case, **0.4625**. This is the area between the mean and the z score value of 1.78 (above the mean). To this value we must add the area of 0.5000, which is the area below the mean (recall that for a symmetrical normal distribution, the mean divides the distribution into two equal halves, i.e., 0.5000 above and 0.5000 below the mean). Thus, the total area below the z score value of 1.78 is **0.9625** (0.5000 + 0.4625). We can transform this value into a percentage by simply multiplying this score by 100

TABLE 8.1

SPSS Data File Containing the First 10 Computed z Scores

	test_scores	Ztest_scores
1	65.00	−0.16492
2	70.00	0.32014
3	75.00	0.80520
4	55.00	−1.13504
5	85.00	1.77532
6	60.00	−0.64998
7	76.00	0.90221
8	45.00	−2.10516
9	80.00	1.29026
10	64.00	−0.26193

$(0.9625 \times 100 = 96.25\%)$. Thus, it can be concluded that 96.25% of your class of 50 students (or ~48 students) got a lower test score than you.

Going back to our question, "How well did you perform in the statistics exam compared to the other 50 students?" Clearly it can be seen that you did better than a large proportion of the students in your class, with **96.25%** of the class (or ~48 students) scoring lower than you.

8.3.5 Calculating the Percentage of Scores that Fall between Two Known Scores

Using the same sample of 50 statistics exam scores, what percentage of scores fall between 60 and 70? Figure 8.2 presents the relevant diagram.

To solve this problem, we need to (1) find the percentage of scores between the mean and the score of 70, (2) find the percentage of scores between the mean and the score of 60, and (3) add the two percentages of scores together. To do this, we need to compute two z scores that correspond to the two raw scores of 60 and 70. These two computed z scores are presented in Table 8.1. It can be seen from this table that the raw score of 60 has a z score value of **−0.65** (0.65 of a standard deviation below the mean). It can also be seen from this table that the raw score of 70 has a z score value of **0.32** (0.32 of a standard deviation above the mean). Referring to the z score distribution table (see Table A in the Appendix) it can be seen that a z score value of 0.32 (corresponding to the raw score of 70) has a corresponding value of **0.1255**.

That is, **12.55%** of the scores fall between the mean and the raw score of 70. It can also be seen that a z score value of −0.65 (corresponding to the raw score of 60) has a corresponding value of **0.2422**. That is, **24.22%** of the scores fall between the mean and the raw score of 60. Thus, to find the percentage

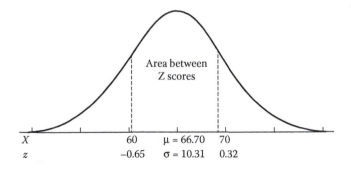

FIGURE 8.2
Area under the normal curve bounded by the scores 60 and 70.

of scores that fall between 60 and 70, simply add **12.55%** (the percentage of scores that fall between the mean and the raw score of 70) to **24.22%** (the percentage of scores that fall between the mean and the raw score of 60).

$$\text{Percentage of scores between 60 and 70} = 12.55\% + 24.22\% = \mathbf{36.77\%}$$

Therefore, 36.77% of scores fall between the raw scores of 60 and 70.

8.3.6 Calculating the Percentile Point with *z* Scores

As described in Chapter 4, a percentile (or percentile point) is a measure indicating the value below which a given percentage of scores in a group of scores fall. This definition is simply the opposite of the percentile rank. Whereas with percentile rank, we are interested in the percentage of scores that fall below a given score (in the above example, 62.55% of the class exam scores fall below the score of 70), with percentile point we are interested in the *value of the score* below which a specified percentage of scores fall. Thus, the 30th percentile is the value (or score) below which 30% of the distribution scores fall.

Let's use an example to demonstrate how we can use *z* scores to calculate the percentile. Using the same data set of 50 statistics exam scores (**EX11. SAV**), what mark would a student have to achieve to be in the top 10% of the class? That is, what is the exam score below which 90% of the class's scores will fall (the 90th percentile)? To answer this question, we need to find the score (X) on our frequency distribution that reflects the top 10% of scores.

First, we convert the frequency distribution of the 50 statistics exam scores into a standard normal distribution. As such, our mean score of 66.70 becomes 0 and the score X we are looking for becomes our *z* score, which is currently unknown. Figure 8.3 presents this standard normal distribution.

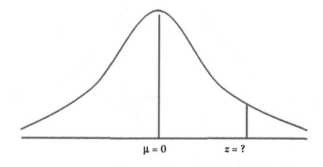

FIGURE 8.3
Standard normal distribution with a mean of 0 and an unknown z score.

Next, we need to calculate the area under the curve that falls below this unknown z score. We know the score (X) a student has to achieve to be in the top 10% of the class is the same score below which 90% (0.90) of the class's scores will fall (i.e., the 90th percentile). Therefore, to calculate this unknown z score, all we need to do is to refer to the z score distribution table that corresponds to 0.4000, the area above the mean (recall that 0.5000 of a normal symmetrical distribution lies below the mean) (see Table A in the Appendix). When looking at the table, it can be seen that the closest value to 0.4000 is 0.3997. Taking this 0.3997 value as our starting point, we can see that the value on the y-axis for 0.3997 is 1.2. We can also see that the value on the x-axis for 0.3997 is 0.08. Putting these two values together, the z score for 0.3997 is **1.28** (i.e., $1.2 + 0.08 = 1.28$). That is, approximately 90% ($50.00\% + 39.97\% = $ **89.97%**) of the class's exam scores fall below the z score of 1.28 (please note that if we use a z score calculator, the value of 0.90 corresponds with a more precise z score of 1.282). Figure 8.4 presents the calculated z score value of 1.282 below which approximately 90% (0.90) of the class's exam scores fall.

We now have the key information (i.e., $\mu = 66.70$, $\sigma = 10.31$, z score $= 1.282$) needed to answer our question directly, namely, What mark (X) would a student have to achieve to be in the top 10% of the class? To compute the mark X, we need to do some basic algebraic manipulation of the z score equation. Recall that the z score equation is

$$z \text{ score} = \frac{X - \mu}{\sigma}$$

To compute the value of X, we cross multiply the parameters on either side of the equal sign. Thus,

$$X - \mu = z(\sigma)$$

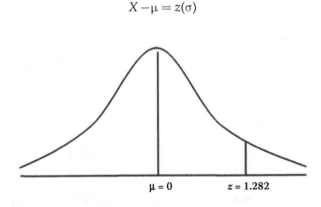

$\mu = 0$ $z = 1.282$

FIGURE 8.4
Standard normal distribution showing the calculated z score value of 1.282 below which 90% (0.90) of the class's exam scores fall.

We then move the mean (μ) to the right side of the equation (the μ becomes positive). Thus,

$$X = z(\sigma) + \mu$$

To solve for X, the mark a student would have to achieve to be in the top 10% of the class, substitute the values of $\mu = 66.70$, $\sigma = 10.31$, and $z = 1.282$ into the equation above. Thus,

$$X = z(\sigma) + \mu$$
$$X = 1.282(10.31) + 66.7$$
$$X = \mathbf{79.92}$$

Therefore, a student must achieve a mark of approximately **80** to be in the top 10% of the class.

8.3.7 SPSS Windows Method

1. Launch the SPSS program and then open the data file **EX11.SAV** (this file contains the 50 statistics exam scores). Click **Analyze** on the menu bar, then **Descriptive Statistics**, and then **Frequencies.** The following **Frequencies** Window will open.

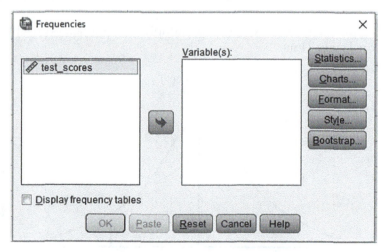

2. In the left-hand field containing the study's **test_scores** variable, click (highlight) this variable, and then click ➡ to transfer the selected **test_scores** variable to the **Variable(s):** field.

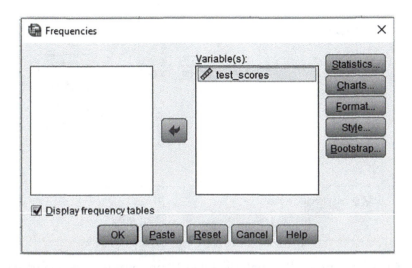

3. Click [Statistics...] to open the **Frequencies Statistics** dialog box. Check the **Percentile(s):** cell and then type **90** in the dialog window (this requests SPSS to calculate the 90th percentile for the set of 50 exam scores); click [Add] to transfer this number to the dialog window below.

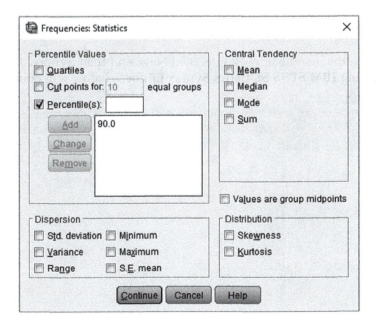

4. Click [Continue] to return to the **Frequencies** window.

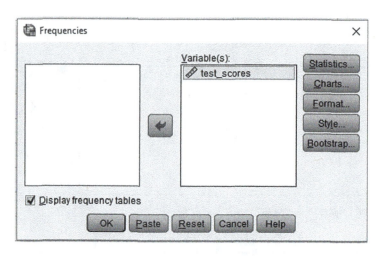

Click [OK] to run the analysis. Table 8.2 presents the exam mark corresponding to the 90th percentile.

8.3.8 SPSS Syntax Method

```
FREQUENCIES VARIABLES=TEST_SCORES
/FORMAT=NOTABLE
/PERCENTILES=90.0
/ORDER=ANALYSIS.
```

1. From the menu bar, click **File**, then **New**, and then **Syntax.** The following **IBM SPSS Statistics Syntax Editor** Window will open.

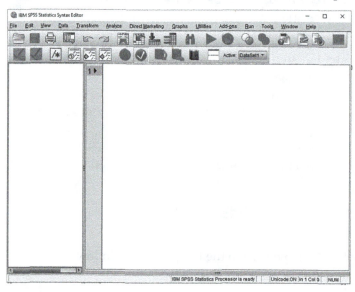

2. Type the **Frequencies** analysis syntax command in the **IBM SPSS Statistics Syntax Editor** Window.

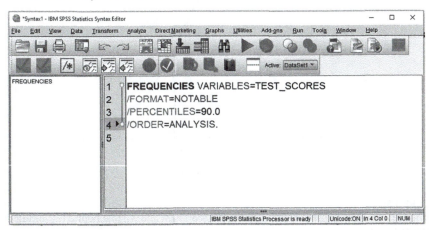

3. To run the **Frequencies** analysis, click ▶ or click **Run** and then **All**.

8.3.9 Table Showing the 90th Percentile for the Set of 50 Exam Scores

It can be seen that the exam mark corresponding to the 90th percentile is approximately 80.

TABLE 8.2

The 90th Percentile for the Set of 50 Exam Scores

Statistics		
test_scores		
N	Valid	50
	Missing	0
Percentiles	90	80.9000

8.3.10 Calculating the Scores that Bound a Specified Area of the Distribution

What are the scores that bound the middle 70% of the statistics exam's distribution? Figure 8.5 presents this diagram.

Figure 8.5 shows the middle 70% of the distribution bounded by the unknown scores X_1 and X_2. Since the distribution is normal and therefore symmetrical, we know that the area between the mean and the score X_1 is 35%; similarly, the area between the mean and the score X_2 is also 35%. As such, the area below the score X_1 is 15%; similarly, the area above the score X_2 is also 15%. To calculate the scores that bound the middle 70% of

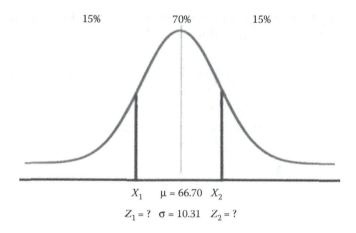

FIGURE 8.5
Normal distribution of exam scores showing the scores (X_1 and X_2) that bound the middle 70% of the distribution.

the statistics exam's distribution, we need to calculate the z score that corresponds to the 35% of the area below the mean (z_1) and the z score that corresponds to the 35% of the area above the mean (z_2). To calculate these unknown z scores, all we need to do is to refer to the z score distribution table that corresponds to 0.3500, the area above and below the mean (see Table A in the Appendix). Looking at the table, it can be seen that the closest value to 0.3500 is 0.3508. Taking this 0.3508 value as our starting point we can see that the value on the y-axis for 0.3508 is 1.0. We can also see that the value on the x-axis for 0.3508 is 0.04. Putting these two values together, the z score for 0.3508 is **1.04** (i.e., 1.0 + 0.04). That is, approximately the middle 70% (35.00% + 35.00% = 70.00%) of the class's exam scores is bounded by the z score values of −1.04 and +1.04. Figure 8.6 presents the calculated z score values of ±1.04 that bound the middle 70% of the class's exam scores.

We now have the key information (i.e., $\mu = 66.70$, $\sigma = 10.31$, z scores $= \pm 1.04$) needed to answer our question directly, namely: What are the scores (X_1 and X_2) that bound the middle 70% of the statistics exam's distribution? To compute the scores X_1 and X_2, apply the equation presented below.

$$X = z(\sigma) + \mu$$

To solve for X_1, the lower bound score for the 70% distribution, substitute the values of $\mu = 66.70$, $\sigma = 10.31$, and $z = -1.04$ into the equation above. Thus,

$$X_1 = z(\sigma) + \mu$$
$$X_1 = -1.04(10.31) + 66.7$$
$$X_1 = \mathbf{55.98}$$

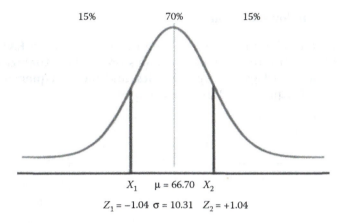

FIGURE 8.6
Normal distribution of exam scores showing the z score values (z_1 and z_2) of $\pm$ 1.04 that bound the middle 70% of the distribution.

To solve for X_2, the upper bound score for the 70% distribution, substitute the values of $\mu = 66.70$, $\sigma = 10.31$, and $z = 1.04$ into the equation above. Thus,

$$X_2 = z(\sigma) + \mu$$
$$X_2 = 1.04(10.31) + 66.7$$
$$X_2 = \mathbf{77.42}$$

Figure 8.7 presents the approximate lower and upper bound scores that bound the middle 70% of the exam distribution.

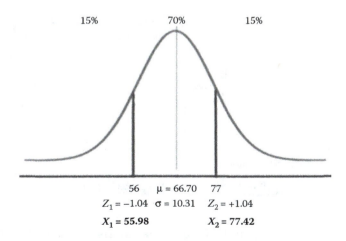

FIGURE 8.7
Normal distribution of exam scores showing the approximate lower and upper bound exam scores (56 and 77) that bound the middle 70% of the distribution.

8.3.11 SPSS Windows Method

1. Launch the SPSS program and then open the data file **EX11.SAV** (this file contains the 50 statistics exam scores). Click **Analyze** on the menu bar, then **Descriptive Statistics**, and then **Frequencies.** The following **Frequencies** Window will open.

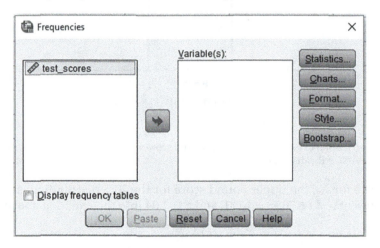

2. In the left-hand field containing the study's **test_scores** variable, click (highlight) this variable, and then click ✦ to transfer the selected **test_scores** variable to the **Variable(s):** field.

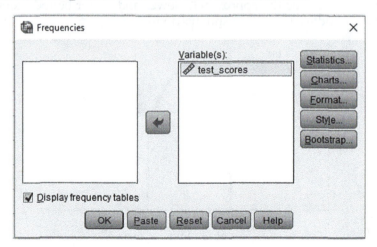

3. Click Statistics... to open the **Frequencies Statistics** dialog box. Check the **Percentile(s):** cell and then type **85** in the dialog window (this requests SPSS to calculate the 85th percentile for the set of 50 exam scores, which corresponds to the upper bound score (X_2) of the 70%

distribution (70% + 15%); click [Add] to transfer this number to the dialog window below.

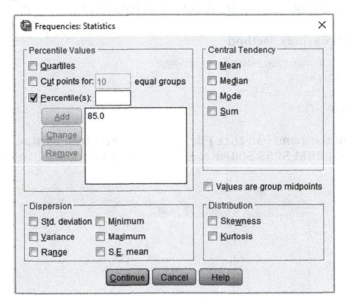

4. Click [Continue] to return to the **Frequencies** Window.

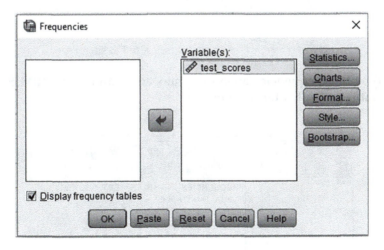

Click [OK] to run the analysis. Table 8.3 presents the exam mark corresponding to the 85th percentile.

To calculate the lower bound score (X_1) of the 70% distribution, repeat Step 1 to Step 4 above. In Step 3, after checking the **Percentile(s):** cell, type **15** in the dialog window (this requests SPSS to calculate the 15th percentile for the set of 50 exam scores that correspond to the lower bound score (X_1) of the 70% distribution (50%–35%). Table 8.3 presents the lower and upper bound scores that bound the middle 70% of the statistics exam's distribution.

8.3.12 SPSS Syntax Method

```
FREQUENCIES VARIABLES=TEST_SCORES
/FORMAT=NOTABLE
/PERCENTILES=85.0 15.0
/ORDER=ANALYSIS.
```

1. From the menu bar, click **File**, then **New**, and then **Syntax.** The following **IBM SPSS Statistics Syntax Editor** Window will open.

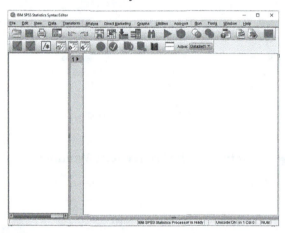

2. Type the **Frequencies** analysis syntax command in the **IBM SPSS Statistics Syntax Editor** Window.

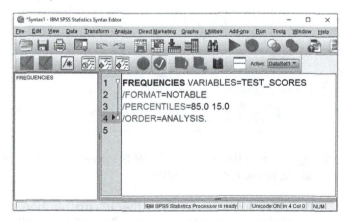

3. To run the **Frequencies** analysis, click ▶ or click Run and then **All.**

8.3.13 Table from either Window or Syntax Methods for Displaying Lower and Upper Bound Scores Binding the Middle 70% of the EX11.SAV data set

TABLE 8.3

The Approximate Lower and Upper Bound Scores that Bound the Middle 70% of the Statistics Exam Distribution

Statistics		
test_scores		
N	Valid	50
	Missing	0
Percentiles	15	56.3000
	85	76.0000

It can be seen that the lower and upper bound exam scores that bound the middle 70% of the statistics exam distribution are approximately 56 and 76, respectively.

8.3.14 Using z Scores to Compare Performance between Different Distributions

The above examples show that, in conjunction with the normal curve, z scores allow us to determine the *percentile rank* of any score in the distribution (the percentage of scores that fall below a specified score in the distribution) as well as the *percentile point* (the score below which a specified percentage of scores fall). In addition, z scores allow comparison between scores in different distributions, even when the units of measurement in the distributions are different. This ability to make meaningful comparisons between scores with different distributions is especially beneficial to educators because it allows comparisons to be made between tests with different distribution characteristics, that is, mean and standard deviation. Let's use an example to demonstrate this.

In addition to sitting the statistics exam, your class of 50 students also sat a chemistry exam. Recall that your statistics exam mark is 85. For your chemistry exam, you scored a mark of 80. Since your statistics exam mark is higher than your chemistry exam mark, you might think that you are better in statistics than in chemistry. However, this is not always true. You can only say that you are better in a particular subject if you get a score with a certain number of standard deviations above the mean. Let's take a closer look at the characteristics of your statistics and chemistry exam scores. These characteristics clearly show that the two set of exam scores have different distributions (different means and standard deviations).

Statistics Exam	Chemistry Exam
Mean (μ) = 66.70	Mean (μ) = 53.90
SD (σ) = 10.31	SD (σ) = 12.22
X_1 (exam mark) = 85.00	X_2 (exam mark) = 80.00
z score = 1.78	z score = 2.14

Your statistics exam mark of 85 has a transformed z score value of 1.78, which means that your statistics exam mark is 1.78 standard deviations above the mean. Your chemistry exam mark of 80 has a transformed z score value of 2.14, which means that your chemistry exam mark is 2.14 standard deviations above the mean. Using these z scores to determine their corresponding percentile ranks, it can be seen that your statistics exam mark of 85 has a corresponding percentile rank of 0.9625 (see Table A in the Appendix); therefore, you scored better than 96.25% or approximately 48 students in your class. It can also be seen that your chemistry exam mark of 80 has a corresponding percentile rank of 0.9838 (see Table A in the Appendix); therefore, you scored better than 98.38% or approximately 49 students in your class. Based on this data, you actually performed better in chemistry than in statistics!

Let's do another example. Let's say we measured the height and weight of the 50 students in your class. The characteristics of these height and weight measurements are as follows:

Height (in cm)	Weight (in kg)
Mean (μ) = 160 cm	Mean (μ) = 80 kg
SD (σ) = 5 cm	SD (σ) = 20 kg
Your height = 170 cm	Your weight = 120 kg

What are the percentile ranks corresponding to your height of 170 cm and your weight of 120 kg? In determining the percentile ranks, we need to first convert the raw scores of 170 cm and 120 kg to their corresponding z scores.

$$z \text{ score} = \frac{X - \mu}{\sigma}$$

Height

$$z \text{ score} = \frac{X - \mu}{\sigma} = \frac{170 - 160}{5} = 2.00$$

Weight

$$z \text{ score} = \frac{X - \mu}{\sigma} = \frac{120 - 80}{20} = 2.00$$

Therefore, your height of 170 cm and your weight of 120 kg have identical z score values of 2.00. Refer to Table A in the Appendix to determine their corresponding percentile ranks. From this z score table, it can be seen that a z score of 2.00 (2 standard deviations above the mean) has a corresponding percentile rank of 97.72% (50.00% + 47.72%). Therefore, both your height and your weight have the same percentile rank of 97.72%. It seems that your height of 170 cm and your weight of 120 kg have something in common. Since they have the same percentile rank, they occupy the same relative position in their respective distributions. As strange as it may sound, you are as heavy as you are tall!

The above two examples demonstrate a very important use of z scores – they allow us to compare scores that are not normally comparable. For example, a person's height and weight cannot be directly compared as they are measured on different scales (centimetres and kilograms). However, when we convert the raw scores into z scores, we have in effect eliminated the original units of measurement (centimetres and kilograms) and replaced them with a universal unit, the standard deviation. Thus, your height of 170 cm and your weight of 120 kg become the same transformed z score value of 2 standard deviation units above the mean, with the same corresponding percentile rank of 97.72%. As can be seen, converting raw scores into their corresponding z scores allows us to compare 'apples with oranges', as long as their distributions permit computation of their means and standard deviations.

9

Correlation

9.1 The Concept of Correlation

In the previous chapters, we were mainly interested in single variables and how to calculate various statistics that describe the distribution of the values of these variables. For example, frequency distributions, measures of central tendency and variability, the normal curve and z scores, all describe the characteristics of single variables and how to relate these statistics to the interpretation of individual scores. However, many of the questions raised in the behavioral sciences go beyond the description of a single variable. We are often called upon to determine whether scores of one variable are related to the scores of another variable. For example, university administrators are often concerned with the relationship between grades earned by high school students and their performance at university. A question that is often asked is *"do students who do well in high school also perform well at university?"* This question reflects the administrators' interest in knowing whether there is a relationship between high school grades earned by the students and their success at university. If a strong relationship between these two variables did exist, then high school grades could be used to predict success in university and therefore would be useful in screening prospective university students. There are many other questions that can be addressed by investigating the relationship between variables. Do parents who suffer from mental health problems tend to have children who are also prone to suffer from mental health problems? Do people who smoke on a regular basis have a higher tendency to contract lung cancer? Is there a relationship between gasoline conservation advertising campaigns and reduction in gasoline usage per month? When we raise questions about the relationships among variables, we are concerned with the topic of correlation.

9.2 Linear and Nonlinear Relationships

The first thing to recognize with correlation is that it works best when the relationship between two variables is *linear*. A linear relationship between

TABLE 9.1

Soft Drink Sales and Temperature on that Day

Temperature °C (X)	Soft Drink Sales (Y)
14.2°	$215
16.4°	$325
11.9°	$185
15.2°	$332
18.5°	$406
22.1°	$522
19.4°	$412
25.1°	$614
23.4°	$544
18.1°	$421
22.6°	$445
17.2°	$408

two variables is one in which the relationship can best be represented by a straight line. The following example illustrates this.

The local grocery store is interested in knowing whether a relationship exists between the amount of soft drinks it sells and the temperature on that day. The storeowner recorded the figures for the last 12 days. Table 9.1 presents these figures.

Figure 9.1 presents these data as a *Scatter Plot* (a scatter plot has points that show the relationship between two sets of data (*XY*)).

We can see from the scatter plot that the relationship between day temperature and soft drink sales is represented by a straight line going from bottom left to upper right. This relationship is therefore linear and informs the storeowner that warmer weather leads to more sales.

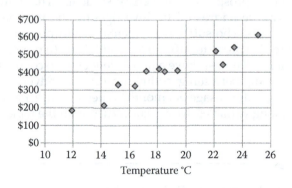

FIGURE 9.1

Scatter plot of linear relationship between soft drink sales and temperature on that day.

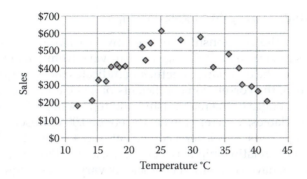

FIGURE 9.2
Scatter plot of nonlinear relationship between soft drink sales and temperature on that day.

The correlation calculation works well for the above example as the relationship between the two variables is linear. The calculation does not work well for nonlinear relationships. This can be demonstrated from an extension of the above example. Suppose there has been a heat wave and the temperature has soared. It has gotten so hot that people have opted to stay at home rather than go out. As a result, the soft drink sales at the grocery store start dropping. The following graph (Figure 9.2) shows sales decreasing as the temperature increased.

We can see from the graph that soft drink sales did increase in a linear fashion, at least up to a maximum of 25°C. However, as the temperature continued to rise, sales began decreasing, resulting in an overall nonlinear curve. Correlation calculation for this set of 'nonlinear' data will return a correlation value of 0, which means 'no relationship' between the two variables. However, inspection of the graph clearly shows that the data set does have a correlation; it follows a linear relationship that reaches a peak at around 25°C. In such a case, a scatter plot diagram can roughly indicate the existence or otherwise of a nonlinear relationship.

9.3 Characteristics of Correlation

A correlation is a single number that describes the characteristics of the relationship between two variables. These characteristics concern:

1. The magnitude of the relationship, that is the strength of the relationship
2. The direction of the relationship, that is, whether relationship is positive or negative

9.3.1 Magnitude (Strength) of Relationships

The strength of the relationship between two variables can be expressed statistically as a *correlation coefficient* and can vary from +1.00 to −1.00. The numerical part of the correlation coefficient indicates the strength of the relationship. Indeed, the correlation coefficients of +1.00 and −1.00 are extreme indices of the strength of the relationship and represent perfect relationships between the variables; a correlation coefficient of 0.00 represents the absence of a relationship. Thus, the closer a coefficient is to +1.0 or −1.0, the greater is the strength of the relationship between the variables. Please note that *imperfect* relationships have correlation coefficients varying in magnitude from 0 to +1.00. The sign (+ or −) before the number indicates the direction of the relationship.

9.3.2 Direction of Relationships

The direction of the relationship between any two variables can be *positive* or *negative*. As mentioned above, a correlation coefficient can vary from +1.00 to −1.00. While the numerical part of the correlation coefficient indicates the strength of the relationship, the sign (+ or −) before the number indicates the direction of the relationship.

1. A *positive* (+) *relationship* means that individuals obtaining high scores on one variable tend to obtain high scores on a second variable. That is, as one variable increases in magnitude, the second variable also increases in magnitude. The converse is equally true, that is, individuals scoring low on one variable tend to score low on a second variable. The following graph shows a positive relationship between the two variables X and Y. For a positive relationship, the graph line runs upward from bottom left to upper right.

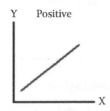

2. A *negative* (−) *relationship* (sometimes called an *inverse relationship*) means that individuals scoring low on one variable tend to score high on a second variable. Conversely, individuals scoring high on one variable tend to score low on a second variable. The following graph shows a negative/inverse relationship between the two variables X and Y. For a negative/inverse relationship, the graph line runs downward from upper left to bottom right.

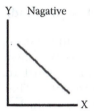

9.4 Correlation Coefficient and z Scores

A major problem in measuring the relationship between two variables is that very often the variables are measured on different scales and in different units. For example, if we are interested in measuring the relationship between soft drink sales and temperature on that day for the data presented in Table 9.1, we face the problem that these two variables are measured on different scales ($ vs. Celsius). However, as was demonstrated in Chapter 8, we can solve this problem by converting the raw scores ($ and Celsius) into z scores. Recall that by transforming the two raw scores into z scores, we have in effect put both variables on the same z 'yardstick'. Putting both variables on the same z scale enables the meaningful measurement of their relationship.

Let's take an example to demonstrate the usefulness of z scores in computing the correlation between two variables that are measured on different scales. Suppose you measured the weight and height of 6 people and you want to know whether there is a relationship between these two variables. A simple scatter plot of these two variables will show whether a relationship exist. Table 9.2 presents data for the weight (kg) and height (cm) of the 6 persons.

TABLE 9.2

Weight (kg) and Height (cm) of 6 Persons

Weight (kg)	Height (cm)
52.25	125
53.00	150
53.75	175
54.50	200
55.25	225
56.00	250

9.4.1 Scatter Plot (SPSS Windows Method)

1. Launch the SPSS program and then open the data file **EX12.SAV** (this file contains the 6 pairs of measurements for the weight and height variables). From the menu bar, click **Graphs**, then **Legacy Dialogs**, and then **Scatter/Dot...** The following **Scatter/Dot** window will open. Click (highlight) the icon.

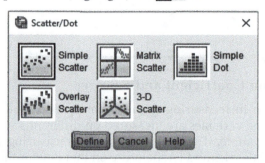

2. Click to open the **Simple Scatterplot** Window below.

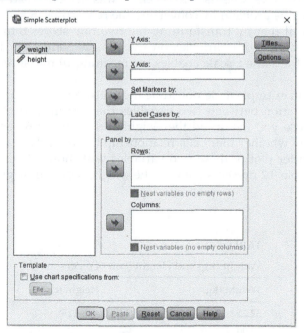

3. Transfer the **WEIGHT** variable to the **Y Axis:** field by clicking (highlight) the variable and then clicking . Transfer the **HEIGHT** variable to the **X Axis:** field by clicking (highlight) the variable and then clicking . Click to complete the analysis. See Figure 9.3 for the scatter plot.

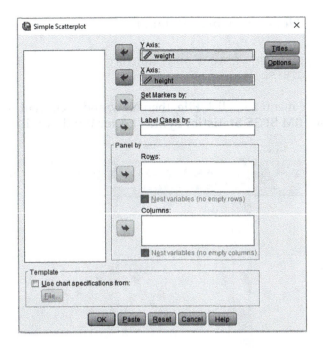

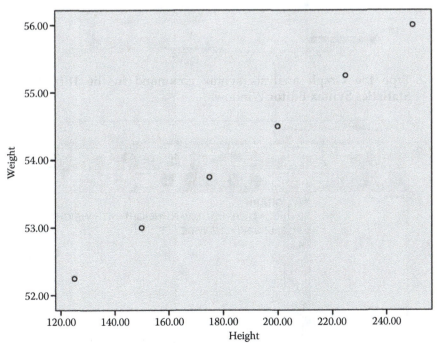

FIGURE 9.3
Scatter plot of the relationship between the variables weight and height.

9.4.2 Scatter Plot (SPSS Syntax Method)

```
GRAPH
/SCATTERPLOT(BIVAR)=HEIGHT WITH WEIGHT
/MISSING=LISTWISE.
```

1. From the menu bar, click **File**, then **New**, and then **Syntax.** The following **IBM SPSS Statistics Syntax Editor** Window will open.

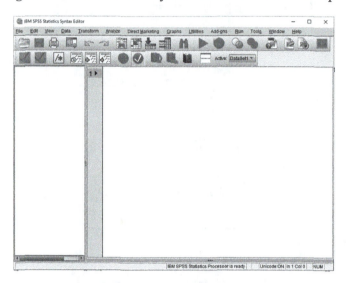

2. Type the **Graph** analysis syntax command in the **IBM SPSS Statistics Syntax Editor** Window.

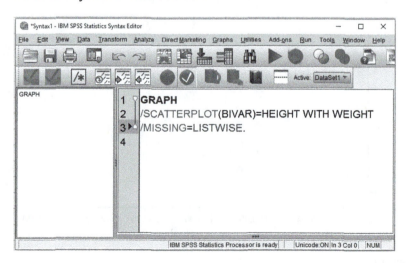

3. To run the **Graph** analysis, click ▶ or click Run and then **All**.

9.4.3 Scatter Plot

Are the two variables of weight and height for the 6 persons related? Looking at the scatter plot for these two variables (Figure 9.3), the answer is definitely 'yes'. All the points fall on a straight line going from bottom left to upper right. These characteristics show that there is a *perfect positive correlation* between the weight and height of the 6 persons. As such, the correlation coefficient must equal +1.00. However, this perfect correlation between the two variables of weight and height is not obvious from their raw scores as the relationship is obscured by the differences in scaling between the two variables ($ vs. Celsius). In order to make the two measurements comparable, we have to convert the raw scores into z scores.

9.4.4 Converting Raw Scores into z Scores (SPSS Windows Method)

1. Launch the SPSS program and then open the data file **EX12.SAV**. Click **Analyze** on the menu bar, then **Descriptive Statistics** and then **Descriptives**. The following **Descriptives** window will open. Check the **Save standardized values as variables** cell.

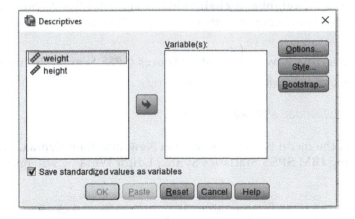

TABLE 9.3

SPSS Data File Containing the 6 Pairs of Raw Scores and Their Corresponding Computed z Scores

	weight	height	Zweight	Zheight
1	52.25	125.00	−1.33631	−1.33631
2	53.00	150.00	−0.80178	−0.80178
3	53.75	175.00	−0.26726	−0.26726
4	54.50	200.00	0.26726	0.26726
5	55.25	225.00	0.80178	0.80178
6	56.00	250.00	1.33631	1.33631

2. In the left-hand field containing the study's **weight** and **height** variables, click (highlight) these variables, and then click ![transfer button] to transfer the selected **weight** and **height** variables to the **Variable(s):** field.

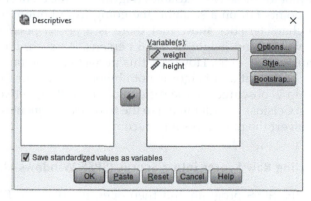

3. Click ![OK] to run the analysis. In running this analysis, SPSS will transform the 6 pairs of raw scores into their corresponding z (standard) scores and will append these z scores as new variables with the name **Zweight** and **Zheight** in the data set. Table 9.3 presents the 6 pairs of raw scores and their corresponding computed z scores.

9.4.5 Converting Raw Scores into z Scores (SPSS Syntax Method)

```
DESCRIPTIVES VARIABLES=WEIGHT HEIGHT
/SAVE
/STATISTICS=MEAN STDDEV MIN MAX.
```

1. From the menu bar, click **File**, then **New**, and then **Syntax.** The following **IBM SPSS Statistics Syntax Editor** Window will open.

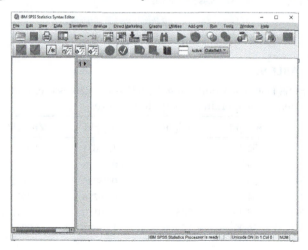

2. Type the **Descriptives** analysis syntax command in the **IBM SPSS Statistics Syntax Editor** Window.

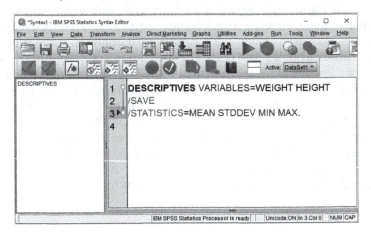

3. To run the **Descriptives** analysis, click ▶ or click Run and then **All**.

9.4.6 SPSS Data File Containing the 6 Pairs of Computed *z* Scores

From the above table, it can be seen that each pair of raw scores now have their identical corresponding pair of *z* scores. For example, for the first person whose weight is 52.25 kg and height is 125 cm, their corresponding transformed *z* score values are now the same, that is, −1.33631. Thus, transformation 'equates' the original raw scores of 52.25 (weight) and 125.00 (height) by placing both measurements on the same *z* scale such that they are both 1.33631 standard deviation units below the means in their respective distributions. The same is true for all the other pairs of raw scores. After transformation into *z* scores, each pair of raw scores occupies the same relative position within their own distributions. Therefore, their corresponding pair of *z* score values is identical. If each pair of *z* scores has the same value, then the correlation between the two identical *z* distributions (**Zweight** and **Zheight**) must be perfect, that is, +1.00. This perfect correlation is not obvious from the raw scores obtained from the weight and height distributions.

9.5 Pearson *r* and the Linear Correlation Coefficient

The *Pearson product-moment correlation coefficient* is a measure of the strength and direction of the linear relationship between two variables. It is sometimes referred to as *Pearson's correlation, Pearson r,* or simply as the *correlation coefficient*. As mentioned earlier, in calculating the relationship between two variables, the variables are often measured in different units and have different scaling. The previous example in which we investigated the relationship between the weight and height of a sample of 6 people demonstrates this point. Both these variables have different units of measurement with weight being measured in kilograms and height being measured in centimetres. As such, to conduct a meaningful analysis of the relationship between these variables, the resultant correlation coefficient must be independent of the units of measurement and scaling between the two variables. The Pearson *r* achieves this by transforming the original measurements (kg and cm) into standardized *z* scores. That is, it replaces the original scores with their transformed *z* scores – in effect, putting the measurements of the two variables on the same *z* scale. Thus, in order to calculate the relationship between any two variables, the Pearson *r* equation correlates the variables' *z* scores and not their original raw scores.

The equation for calculating the Pearson *r* is as follows:

$$r = \frac{\Sigma XY - \frac{(\Sigma X)(\Sigma Y)}{N}}{\sqrt{\left[\Sigma X^2 - \frac{(\Sigma X)^2}{N}\right]\left[\Sigma Y^2 - \frac{(\Sigma Y)^2}{N}\right]}}$$

9.5.1 Example of the Pearson *r* Calculation

Assume that an experimenter wishes to determine whether there is a relationship between the GPAs and the scores on a reading-comprehension test of 10 first-year students. Table 9.4 presents these scores.

$$r = \frac{\Sigma XY - \frac{(\Sigma X)(\Sigma Y)}{N}}{\sqrt{\left[\Sigma X^2 - \frac{(\Sigma X)^2}{N}\right]\left[\Sigma Y^2 - \frac{(\Sigma Y)^2}{N}\right]}}$$

$$r = \frac{1201.80 - \frac{(448)(25.40)}{10}}{\sqrt{\left[21146 - \frac{(448)^2}{10}\right]\left[68.98 - \frac{(25.40)^2}{10}\right]}}$$

$$r = \frac{1201.80 - 1137.92}{\sqrt{[21146 - 20070.4][68.98 - 64.516]}}$$

$$r = \frac{63.88}{\sqrt{[1075.6][4.464]}}$$

$$r = \frac{63.88}{69.29} = 0.922$$

TABLE 9.4

GPAs and Scores on a Reading-Comprehension Test of 10 First-Year Students

Students	Reading Score (X)	GPA (Y)	X^2	Y^2	XY
s1	38	2.1	1444	4.41	79.80
s2	54	2.9	2916	8.41	156.60
s3	43	3.0	1849	9.00	129.00
s4	45	2.3	2025	5.29	103.50
s5	50	2.6	2500	6.76	130.00
s6	61	3.7	3721	13.69	225.70
s7	57	3.2	3249	10.24	182.40
s8	25	1.3	625	1.69	32.50
s9	36	1.8	1296	3.24	64.80
s10	39	2.5	1521	6.25	97.50
Total (Σ)	448	25.40	21146	68.98	1201.80

The calculated Pearson r value is 0.922. This indicates that there is a very strong positive relationship between the students' reading-comprehension test scores and their GPAs. Thus, as the students' reading-comprehension test scores increased, so did their GPAs, and vice versa.

9.5.2 SPSS Windows Method

1. Launch the SPSS program and then open the data file **EX13.SAV** (this file contains the 10 pairs of measurements for the reading-comprehension and GPA variables). From the menu bar, click **Correlate**, and then **Bivariate...** The following **Bivariate Correlations** Window will open. Under **Correlation Coefficients** check the **Pearson** cell. Under **Test of Significance**, check the **Two-tailed** cell.

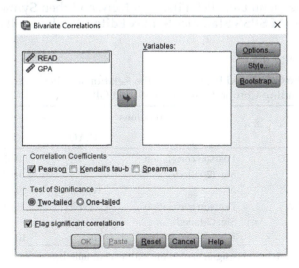

2. Transfer the **READ and GPA** variables to the **Variables:** field by clicking (highlight) the variables and then clicking ⬇. Click ⬜ OK to complete the analysis. See Table 9.5 for the Pearson *r*.

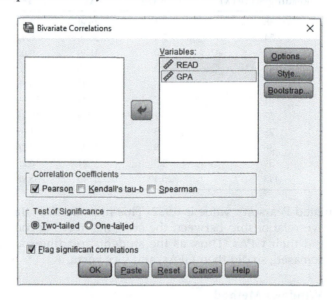

9.5.3 SPSS Syntax Method

```
CORRELATIONS
/VARIABLES=READ GPA
/PRINT=TWOTAIL NOSIG
/MISSING=PAIRWISE.
```

1. From the menu bar, click **File**, then **New**, and then **Syntax.** The following **IBM SPSS Statistics Syntax Editor** Window will open.

TABLE 9.5

The Calculated Pearson *r* for the Relationship between Reading-Comprehension Scores and GPAs

	Correlations		
		READ	**GPA**
READ	Pearson correlation	1	0.922[a]
	Sig. (2-tailed)		0.000
	N	10	10
GPA	Pearson correlation	0.922[a]	1
	Sig. (2-tailed)	0.000	
	N	10	10

[a] Correlation is significant at the 0.01 level (2-tailed).

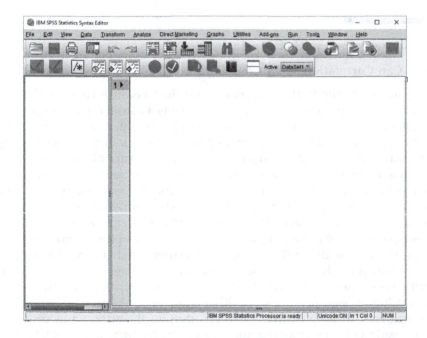

2. Type the **Correlations** analysis syntax command in the **IBM SPSS Statistics Syntax Editor** Window. To run the **Correlation** analysis, click ▶ or click Run and then **All**.

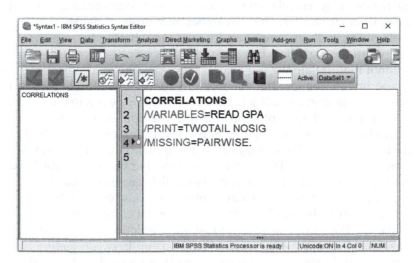

9.5.4 The Calculated Pearson *r*

The SPSS computed Pearson *r* value of **0.922** is identical to the value calculated manually from the Pearson *r* equation in Section 9.5.1.

9.6 Some Issues with Correlation

9.6.1 Can Correlation Show Causality?

The simple answer to this question is no. Just because two variables are correlated does not mean that one variable has a causal effect on the other. Correlation simply means that the variables of interest co-vary, that is changes in one variable are associated with changes in the other variable. But covariation/association does not mean that one variable *causes* the variation in another variable. For instance, when people suffer from colds, they often experience runny noses and sore throats, that is, these two maladies correlate with each other in that they tend to show up simultaneously and in the same patients. However, this does not mean that runny noses cause sore throats, or that sore throats cause runny noses. Let's look at another example, which shows how difficult it is to use correlational data to nail down causation conclusively. In attempts to demonstrate that smoking causes cancer, health researchers provided evidence based on correlational studies that show a highly significant and positive relationship between smoking and the incidence of lung cancer. However, despite the strong and consistent relationship between smoking and cancer, the findings are correlational, a point emphasized by tobacco company lawyers who argued for 40 years that smoking merely "correlated" to lung cancer rather than actually caused it. In other words, correlation does not imply causality.

The reason why correlational studies cannot demonstrate causality is because, in such studies, the researcher measures variables that already naturally exist, such as weight and IQ, to see whether there is a relationship between the variables. The researcher has no control over the variables and cannot manipulate the variables in any way like in a controlled laboratory experiment. The fact that laboratory experiments, unlike correlational studies, can establish causality is because laboratory experiments allow the researcher a high degree of control in the manipulation of the IV and to measure the changes it may have on the DV. By manipulating the IV (e.g., gender), they can see the changes this manipulation (male vs. female) may have on the DV (e.g., IQ). This allows the researcher to compare the IQ scores between the two gender groups and to establish cause and effect.

Therefore, if correlations cannot establish cause and effect, why are they so useful? To answer this question, let's look at what correlations can do. First, correlations allow the researcher to make predictions. For example, past studies have found a strong positive relationship between feeling burnout (exhaustion) and procrastination (putting off one's work). That is, the more exhausted we feel, the higher is our tendency to put off our work. While this finding does not allow the researcher to say whether it is burnout that causes procrastination or vice versa, it does allow counsellors, clinicians, etc. to spot the warning signs. If they know that feeling exhausted at work is linked to the tendency to defer work, then they can potentially prevent it from occurring.

Second, correlations are useful in exploratory research by providing a starting point in areas that have not been previously studied. Through correlational studies, the researcher can initially establish whether there is a relationship between the two variables of interest and, if there is, lead onto a controlled experimental study that can establish whether one variable causes the other. However, there are obvious ethical limits to controlled studies. For example, if you wanted to study whether there is a relationship between smoking and lung cancer, it would be problematic to take two comparable groups and make one group smoke while denying cigarettes to the other in order to see whether cigarette smoking really causes lung cancer. This is why correlational studies are important as they allow researchers to track the relationship between smoking and the incidence of lung cancer across groups (e.g., male vs. female) and time (e.g., over a period of 20 years).

9.6.2 Spurious Correlation

A spurious correlation is a statistical term that describes a situation in which two variables have no direct connection (correlation), but it is incorrectly assumed they are connected as a result of either coincidence or the presence of a third hidden (confounding) factor. A spurious correlation is also sometimes called an "illusory correlation" because statistics (correlation coefficient) point to a relationship between the variables when in fact there is none. Here are some examples of a spurious correlation.

In Nordic countries such as Denmark, during the winter months, storks would flock to and nest on the roof of houses. After a while, many babies would be born, giving rise to the belief that 'storks bring babies'. In other words, there is a highly significant relationship between the presence of storks during the winter months and the subsequent birth of babies. However, even though this relationship is highly obvious, it is clearly a spurious relationship. The relationship is spurious because it is most likely caused by a third unaccounted-for/confounding variable – *the coldness of the winter months*. During the cold winter months, the storks would fly to and nest on the roofs of houses in order to be warmed by the heat emanating from the roofs and chimneys (e.g., from the fireplaces inside the houses). At the same time, people tend to stay in indoors during the cold winter months, and tend to engage in activity that results in the birth of babies nine months later! So, although there may be a strong relationship between the presence of storks and the birth of babies, the two variables are not causally linked. The relationship is clearly due to the third unaccounted-for/confounding variable of 'coldness'.

Here is another example. If the students in a psychology class who had long hair scored higher in the midterm test than those who had short hair, there would be a correlation between hair length and test scores. This observation may lead to the further assumption that students who wished to improve their grades should let their hair grow. Not many people, however, would believe that such a causal link existed. The real cause might be gender (the confounding variable): that is, women (who usually have longer hair) did better in the test.

A final example. A college student notices that on the days she sleeps in and skips her early classes there are a larger amount of traffic accidents on and around campus. The fact that she thinks her sleeping in is causally linked with more traffic accidents is clearly a spurious correlation. After all, what has sleeping in got to do with the incidence of traffic accidents? Nothing! In fact, what may be actually happening is that a third confounding variable – bad weather – is related to both variables. That is, the reason she sleeps in is because of bad weather, and bad weather tends to cause traffic accidents.

Spurious correlations are common and most people are guilty of making them, no matter how basic they may be. In both an academic environment and in an everyday situation, it is important to be aware of them and to think critically about what the actual relationship is between two seemingly related variables. While they may not be as blatantly incorrect as the connection between sleeping in and traffic accidents, it is important to be aware of them and not to make any important conclusions based on a spurious correlation.

10

Linear Regression

10.1 What Is Linear Regression?

Linear regression is highly similar to correlation in that both deal with the relationship between two variables (X and Y). The major difference is that correlation is concerned with the *direction* (positive or negative) and *magnitude* ($0 \rightarrow \pm 1.00$) of the relationship between X and Y, whereas regression is concerned with using the correlation coefficient for *prediction*, that is, given a specific value of X what is the predicted value of Y?

The ability to predict the value of one variable from the value of another is of particular interest to social scientists. For example, university administrators may be interested in knowing whether knowledge of a potential student's IQ will indicate whether that student will successfully complete a university course. Politicians may be interested in knowing whether knowledge of their constituents' prior voting record can help them make informed guesses concerning their constituents' vote in the coming election. Teachers may be interested in knowing whether knowledge of a student's mathematics aptitude score can help them estimate the quality of his performance in a course in statistics. These questions involve predictions from one variable to another, and psychologists, educators, biologists, sociologists, and economists are constantly being called upon to perform this function.

As regression (prediction) employs the correlation coefficient to predict Y from X, it is clear that the accuracy of the prediction depends on the magnitude of the correlation coefficient. If the relationship is perfect, that is, the correlation coefficient $r = \pm 1.00$, then prediction is easy. Perfect correlation between two variables – X and Y – means that there is a 1-to-1 relationship between the variables, and as such 1 unit of change in X will result in exactly 1 unit of change in Y. Unfortunately, most relationships among variables in the social sciences are not perfect, and for these imperfect relationships (the correlation coefficients lie between 0 to ± 1.00), prediction is more complicated.

10.2 Linear Regression and Imperfect Relationships

Let's use an example to demonstrate how we can make predictions from imperfect relationships. Let's return to the data involving reading scores and GPA scores that were presented in Chapter 9 (Table 9.4). These data have been reproduced in Table 10.1. In this instance, consider that we are interested in predicting GPA scores (Y) from the scores on a reading-comprehension test (X) from 10 first-year students.

10.2.1 Scatter Plot and the Line of Best Fit

Linear regression, that is, predicting Y from X, consists of finding the *best fitting line* that comes closest to all the points on a scatter plot formed by the X (READ) and Y (GPA) variables. So the first step is to generate a scatter plot for these two variables together with the line of best fit.

10.2.2 SPSS Windows Method (Scatter Plot and Line of Best Fit)

1. Launch the SPSS program and then open the data file **EX13.SAV** (this file contains the 10 pairs of measurements for the **READ** and **GPA** variables). From the menu bar, click **Graphs**, then **Legacy Dialogs**, and then **Scatter/Dot...** The following **Scatter/Dot** window will open. Click (highlight) the [icon] Simple Scatter icon.

TABLE 10.1

GPAs and Scores on a Reading-Comprehension Test of 10 First-Year Students

Students	Reading Score (X)	GPA (Y)
s1	38	2.1
s2	54	2.9
s3	43	3.0
s4	45	2.3
s5	50	2.6
s6	61	3.7
s7	57	3.2
s8	25	1.3
s9	36	1.8
s10	39	2.5

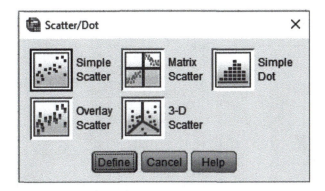

2. Click [Define] to open the **Simple Scatterplot** Window below.

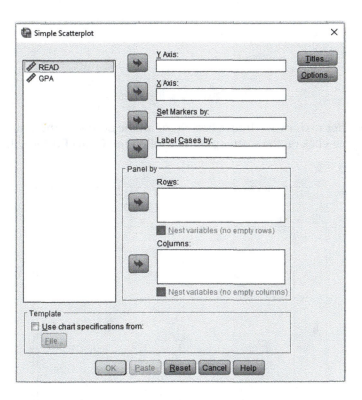

3. Transfer the **GPA** variable to the **Y Axis:** field by clicking (highlight) the variable and then clicking [→]. Transfer the **READ** variable to the **X Axis:** field by clicking (highlight) the variable and then clicking [→]. Click [OK] to complete the analysis. See Figure 10.1 for the scatter plot.

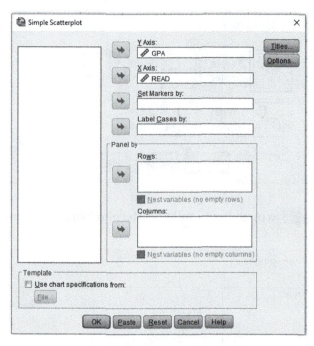

4. In order to draw the line of best fit for all the points on the scatter plot, double-click on the scatter plot. The following **Chart Editor** will open.

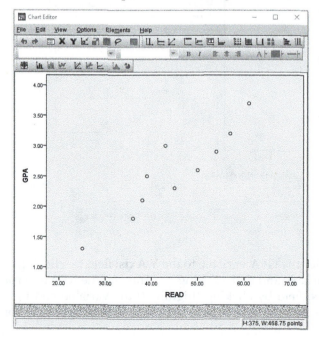

5. Click the 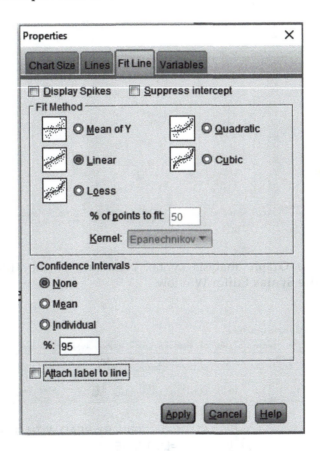 (**Add Fit Line at Total**) icon to add the best fitting line for all the points on the scatter plot. Clicking the ⊾ will also open the **Properties** Window below. Under **Fit Method**, ensure that the **Linear** button is checked. At the bottom of this Window, ensure that the **Attach label to line** cell is unchecked. Click [Apply] to complete the analysis. Finally, close the **Chart Editor** Window. See Figure 10.1 for the scatter plot with its line of best fit.

10.2.3 SPSS Syntax Method (Scatter Plot)

```
GRAPH
/SCATTERPLOT(BIVAR)=READ WITH GPA
/MISSING=LISTWISE.
```

1. From the menu bar, click **File**, then **New**, and then **Syntax.** The following **IBM SPSS Statistics Syntax Editor** Window will open.

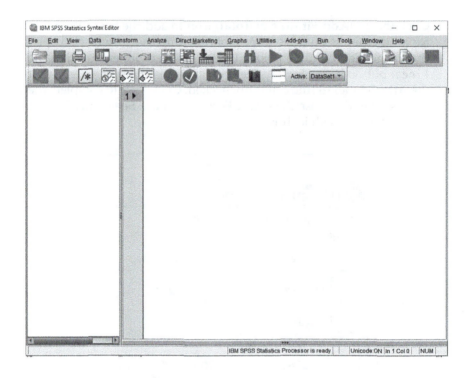

2. Type the **Graph** analysis syntax command in the **IBM SPSS Statistics Syntax Editor** Window.

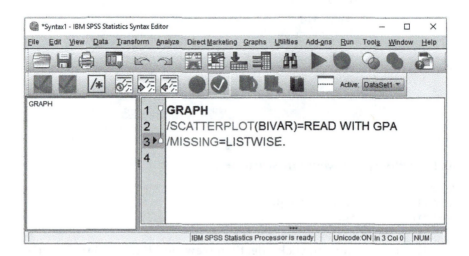

3. To run the **Graph** analysis, click ▶ or click Run and then **All**.

10.2.4 Scatter Plot with Line of Best Fit

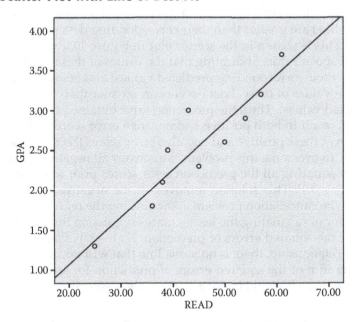

FIGURE 10.1
Scatter plot and line of best fit representing the relationship between the variables READ and GPA.

10.2.5 Least-Squares Regression (Line of Best Fit): Predicting *Y* from *X*

As mentioned earlier, linear regression consists of finding the best fitting line that comes closest to all the points on a scatter plot formed by the *X* (reading scores) and *Y* (GPA) variables. As seen in Figure 10.1, the straight line that runs diagonally from bottom left to upper right is called the *regression line* and represents the predicted score on *Y* for each possible value of *X*. It is also called the *line of best fit* as it comes closest to all the points on the scatter plot. But what is meant by the line of best fit?

The most commonly used description of the line of best fit is the *prediction line that minimizes the total error of prediction, that is, the sum of the squared errors of prediction* [$\Sigma (Y - Y')^2$]. Based on this description, the line of best fit is often called the *least-squares regression line*. Here is the reason why. If we draw a vertical line from each data point on the scatter plot to the regression line, the vertical distance between each data point and the regression line represents the error of prediction. If we let *Y* = the actual data point and *Y'* = the predicted value (the value on the line), then the error of prediction for a data point is the *value of that data point minus the predicted value* (*Y* − *Y'*). Thus, it seems that the total error in prediction is simply the sum of *Y* − *Y'* for all of the data points, that is, $\Sigma (Y - Y')$. As such, if we want to minimize the prediction error, we would construct the regression line that minimizes $\Sigma (Y - Y')$.

There is a problem though. The summation of the prediction error across all the data points does not equal $\Sigma\,(Y - Y')$ because some of the Y' values (on the regression line) are greater than their corresponding data point Y and some are lower. This is shown in the scatter plot in Figure 10.1, where the regression line is above 7 dots (indicating that the values of these 7 data points are lower than their corresponding predicted values) and below 3 dots (indicating that the values of these 3 data points are greater than their corresponding predicted values). Thus, the prediction error obtained from the equation $Y - Y'$ will result in both positive and negative error scores, and the simple summation of these positive and negative error scores $[\Sigma\,(Y - Y')]$ will cancel each other. To overcome this problem, we convert all negative error scores to positive by squaring all the prediction error scores prior to summing them, that is, $\Sigma\,(Y - Y')^2$. This solution removes all the negative error values and eliminates the cancellation problem. Now, finding the regression line of best fit is a matter of calculating the least-squares regression line that minimizes the sum of the squared errors of prediction, $\Sigma\,(Y - Y')^2$. Please note that for any linear relationship, there is only one line that will minimize $\Sigma\,(Y - Y')^2$. That is, the sum of the squared errors of prediction for any linear relationship is lower than it would be for any other regression line.

10.2.6 How to Construct the Least-Squares Regression Line: Predicting Y from X

The equation for the least-squares regression line is

$$Y' = bX + A$$

where
  Y' = predicted value of Y
  b = slope of the line
  $A = Y$ intercept (constant)

The calculations for the equation $Y' = bX + A$ are based on the following statistics:

  $\bar{x}$ = mean of X
  $\bar{Y}$ = mean of Y
  σ_x = standard deviation of X
  σ_y = standard deviation of Y
  r = correlation coefficient between X and Y

The slope (b) can be calculated as

$$b = (r)(\sigma_y)/\sigma_x$$

The Y-intercept/constant (A) can be calculated as

$$A = \bar{Y} - b\bar{x}$$

Applying these statistics to the data set presented in Table 10.1, we get (Table 10.2).

Please note that the calculations of the mean X and Y scores are based on the procedure described in Chapter 6. The calculations of the standard deviations for the X and Y scores are based on the procedure described in Chapter 7. The calculation of the correlation coefficient between X and Y is based on the procedure described in Chapter 9.

To construct the least-squares regression line, we apply the above statistics to the equation $Y' = bX + A$.

The slope (b) can be calculated as follows:

$$b = (r)(\sigma_y)/\sigma_x$$

$$b = (0.922)(0.70427)/10.93211 = \mathbf{0.0593}$$

The Y-intercept (constant) (A) can be calculated as

$$A = \bar{Y} - b\bar{x}$$

$$A = 2.54 - (0.0593)(44.80) = \mathbf{-0.1166}$$

Thus, given that the equation for the least-squares regression line is $Y' = bX + A$ the predicted GPA value (Y') for any value of reading score (X) is

$$Y' = bX + A$$

$$Y' = 0.0593X + (-0.1166)$$

For example, what are the predicted GPA scores for two students who scored 68 and 41 on the reading comprehension test? Based on the above

TABLE 10.2

Means and Standard Deviations of Reading Scores and GPAs and Their Correlation Coefficient

READ Score (X)		GPA (Y)		
$\bar{x}$	σ_x	$\bar{Y}$	σ_y	r
44.80	10.93211	2.54	0.70427	0.922

least-squares regression equation, the predicted GPA score for the student
who scored 68 on the reading test is

$$Y' = 0.0593X + (-0.1166)$$
$$Y' = 0.0593(68) - 0.1166$$
$$Y' = \textbf{3.92}$$

For the student who scored 41 on the reading test, his predicted GPA
score is

$$Y' = 0.0593X + (-0.1166)$$
$$Y' = 0.0593(41) - 0.1166$$
$$Y' = \textbf{2.31}$$

10.2.7 SPSS Windows Method (Constructing the Least-Squares Regression Line Equation)

1. From the menu bar, click **Analyze**, then **Regression**, and then **Linear**
   The following **Linear Regression** Window will open.

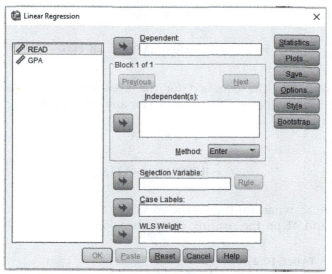

2. Click (highlight) the **GPA** variable and then click ⬛ to transfer this variable to the **Dependent:** field. Next, click (highlight)
   the **READ** variable and then click ⬛ to transfer this variable to the
   Independent(s): field. In the **Method:** field, select **ENTER** from the
   drop-down list as the method of entry for the independent (predictor) variable into the prediction equation.

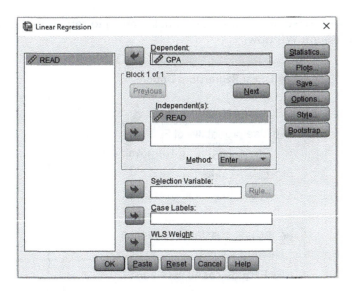

3. Click [Statistics...] to open the **Linear Regression: Statistics** Window. Check the fields to obtain the statistics required. For this example, check the fields for **Estimates, Confidence intervals**, and **Model fit**. Click [Continue] when finished.

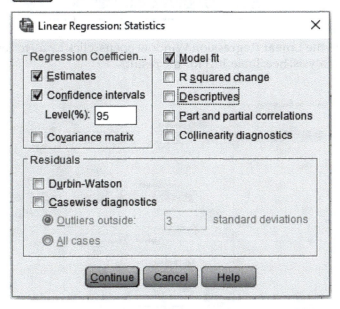

4. When the **Linear Regression** Window opens, click [Options...] to open the **Linear Regression: Options** Window below. Ensure that both the **Use probability of F** and the **Include constant in equation** fields are checked. Under **Missing Values**, ensure that the **Exclude**

cases listwise field is checked. Click ⟨Continue⟩ to return to the **Linear Regression** Window.

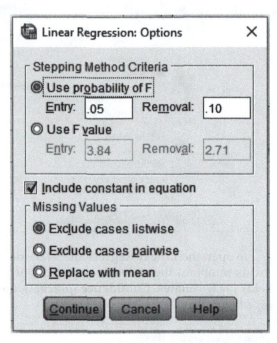

5. When the **Linear Regression** Window opens, click ⟨OK⟩ to complete the analysis. See Table 10.3 for the results.

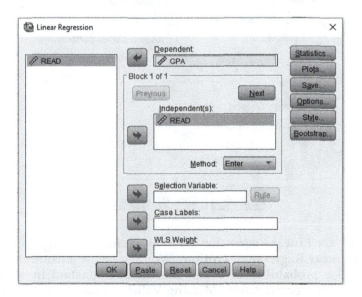

10.2.8 SPSS Syntax Method (Constructing the Least-Squares Regression Line Equation)

```
REGRESSION VARIABLES=(COLLECT)
/MISSING LISTWISE
/STATISTICS=DEFAULTS CI
/DEPENDENT=GPA
/METHOD=ENTER READ.
```

10.2.9 SPSS Output

TABLE 10.3

Linear Regression Analysis Output

	Regression		
	Variables Entered/Removed[a]		
Model	Variables Entered	Variables Removed	Method
1	READ[b]	.	Enter

[a] Dependent variable: GPA.
[b] All requested variables entered.

		Model Summary		
Model	R	R Square	Adjusted R Square	Std. Error of the Estimate
1	0.922[a]	0.850	0.831	0.28943

[a] Predictors: (constant), READ.

		ANOVA[a]				
Model		Sum of Squares	df	Mean Square	F	Sig.
1	Regression	3.794	1	3.794	45.289	0.000[b]
	Residual	0.670	8	0.084		
	Total	4.464	9			

[a] Dependent variable: GPA.
[b] Predictors: (constant), READ.

		Coefficients[a]						
		Unstandardized Coefficients		Standardized Coefficients			95.0% Confidence Interval for B	
Model		B	Std. Error	Beta	t	Sig.	Lower Bound	Upper Bound
1	(Constant)	−0.121	0.406		−0.297	0.774	−1.056	0.815
	READ	0.059	0.009	0.922	6.730	0.000	0.039	0.080

[a] Dependent variable: GPA.

176 *Understanding Statistics for the Social Sciences with IBM SPSS*

10.2.10 Results and Interpretation

1. *Least-squares regression equation*: The relevant information for constructing the least-squares regression equation ($Y' = bX + A$) is presented in the **Coefficients** table (see Table 10.3). Using the **Constant** and **B** values presented in the **Unstandardized Coefficients** column, the least-squares regression equation would be

$$Y'(\text{predicted GPA score}) = (0.059 \times \text{READ}) + (-0.121)$$

   Thus, for a student who has a reading score of 56, his/her predicted GPA score will be

$$\text{Predicted GPA} = (0.059 \times 56) - 0.121$$
$$= \textbf{3.18}$$

   Please note that in the **Model Summary** table, the **Standard Error of the Estimate** is 0.28943. This means that at the 95% confidence interval, the predicted GPA score of 3.18 lies between the scores of **2.61** ($3.18 - (1.96 \times 0.28943)$) and **3.75** ($3.18 + (1.96 \times 0.28943)$).

2. *Evaluating the predictive strength of the least-squares regression equation*: A measure of the strength of the computed equation is **R-square**, sometimes called the **coefficient of determination**. R-square is simply the square of the multiple correlation coefficient listed under **R** in the **Model Summary** table, and represents the proportion of variance accounted for in the DV (GPA) by the predictor variable (READ). In a simple regression such as this, where there is only one predictor variable, the multiple R is equivalent to the simple R (Pearson product-moment correlation). For this example, the multiple correlation coefficient is 0.922, and the R-square is 0.850. Thus, for this sample, the predictor variable of READ has explained 85.0% of the variance in the DV of GPA.

Section II

Inferential Statistics

11

Statistical Inference and Probability

11.1 Introduction to Inferential Statistics

The past 10 chapters dealt specifically with techniques that researchers employ to describe and summarize data in the most economical and meaningful way. Topics such as *frequency distributions, graphing, measures of central tendency, measures of variability, standard (z) scores, correlation,* and *linear regression* all serve this purpose by providing a useful summary of the overall picture of the obtained data set. In presenting research findings, descriptive statistics is very important because if raw data were presented alone, it would be difficult to visualize what the data were showing, especially if there were a lot of it. Descriptive statistics therefore enables the researcher to present the data in a more meaningful way and therefore allows for simpler interpretation of the data. For example, if we had the IQ scores from 100 students, we may be interested in the average IQ of those students (measures of central tendency – mean, median, and mode). We may also be interested in the distribution or spread of the IQ scores (measures of variability – range, standard deviation, and variance). Descriptive statistics allows us to do this.

As useful as descriptive statistics may be, it does not, however, allow the researcher to make conclusions beyond the data analysed or to reach conclusions regarding any hypotheses the researcher might have made. It is simply a way to describe data. In conducting research though, a major aim is clearly to move beyond the mere description of the collected data to the drawing of inferences/conclusions about a population based on information contained in a sample taken from that population. With inferential statistics, the researcher is trying to draw conclusions that extend beyond the immediate data. Thus, when conducting research, the researcher employs inferential statistics to serve two major purposes: (1) *to infer from the sample data the attitudes/opinions of the population* and (2) *hypothesis testing, that is, to aid the researcher in making a decision regarding, for example, whether an observed difference between groups (e.g., gender difference in problem-solving ability) is a meaningful/reliable difference or one that might have happened by chance alone.* Thus, whereas descriptive statistics is used to simply describe what is going on in a data set, inferential statistics is used to infer from the data set to more general conclusions. The topics of *probability* and *random sampling* are crucial to the methodology of inferential statistics.

11.2 Probability

Probability is concerned with the possible chance outcomes of events. In calculating probabilities, that is, assigning relative frequencies to each of the possible outcomes, we can approach the calculations in two ways – the classical approach and the empirical approach.

11.2.1 The Classical Approach to Probability

This approach to the calculation of probability is based on reason/logic alone. For example, if I flip an unbiased coin once, what is the probability that it will turn up heads? We know that there are only two possible outcomes – heads or tails. Thus, the probability that the coin will turn up heads, $p(H)$, is ½ or 0.50. This classical way of deducing probability from logic alone without recourse to empirical data is defined as

$$p(A) = \frac{\text{Number of events favoring A}}{\text{Total number of events}}$$

The above equation tells us that the probability of event A occurring is equal to the number of events favoring A divided by the total number of events. For the coin example, the total number of possible events is 2 (heads or tails), and the relative frequency of each outcome is 1. Thus, $p(H) = ½$ and $p(T) = ½$.

Let's do a couple more examples. If I roll a die (singular for dice) once, what is the probability that it will turn up a 5 (i.e., the side with 5 spots on it)? Since a die has 6 sides and there is only 1 side with 5 spots on it, the probability of getting a 5 from one roll of the die is

$$p(5) = \frac{\text{Number of events favoring 5}}{\text{Total number of events is 6}} = \frac{1}{6} \quad \text{or} \quad 0.1666$$

If I roll a die once, what is the probability that it will turn up a 2 or a 5? Since there are two possible outcomes, the probability of getting a 2 or a 5 from one roll of the die is

$$p(2 \text{ or } 5) = \frac{\text{Number of events favoring 2 or 5}}{\text{Total number of events is 6}} = \frac{2}{6} \quad \text{or} \quad 0.3333$$

11.2.2 The Empirical Approach to Probability

In the above 'classical approach' examples, the probabilities of occurrence were calculated on the basis of logic alone with no recourse to any data collection. This approach is contrasted with the empirical approach in which the calculation of the probability of the occurrence of an event is based on empirical data. This empirical approach to probability is defined as

$$p(A) = \frac{\text{Frequency of A occurring}}{\text{Total number of occurrences}}$$

What is the probability of rolling a 6 from one roll of a die? In order to use the empirical approach to determine the probability of rolling a 6 from 1 roll of a die, we need to take an actual die and roll it many, many times and then calculate the number of times (frequency) that 6 has appeared. The more times we roll the die, the more accurate the computed probability will be. Suppose we roll the die a 1,000,000 times and that 6 appears 160,000 times. Using the 'empirical' equation above, we determine the probability of rolling a 6 from 1 roll of a die as

$$p(6) = \frac{\text{Frequency of A occurring}}{\text{Total number of occurrences}} = \frac{160,000}{1,000,000} = 0.1600$$

Here's another example. If I select randomly one student from a university population of 1,000 students, what is the probability that that student will be blue-eyed? If we found that 300 of the 1,000 students have blue eyes, then we can use the above empirical equation to calculate the probability that the selected student will have blue eyes.

$$p(\text{blue-eyed}) = \frac{\text{Frequency of blue-eyed students}}{\text{Total number of students}} = \frac{300}{1,000} = 0.30$$

Thus, the probability that the selected student will have blue eyes is 0.30. Note that for the above examples, it is necessary to collect some data before determining the probability.

11.2.3 Expressing Probability Values

Probability is expressed numerically as a number between 0 and 1. A probability of 0 means that the event is certain not to occur. For example, what is the probability of rolling a 7 from 1 roll of a die? Since an ordinary die has only 6 sides ranging from 1 spot to 6 spots, rolling a 7 cannot occur. Therefore, the probability of rolling a 7 from 1 roll of a die is 0. A probability of 1 means that the event will happen. What is the probability of getting a head or a tail from one flip of a coin? Since a coin has only two sides – heads and tails – it is certain that either a head or a tail will turn up. Therefore, the probability of getting a head or a tail from one flip of a coin is 1.

The probability of an event occurring can be expressed as a fraction, a decimal point, a percentage, chances in 100, or as odds for or against. For example, the probability of rolling a 5 from 1 roll of a die can be expressed as 1/6 or 0.1666 or 16.66%. It can also be expressed as having 16.66 chances in 100 of rolling a 5. In betting circles, it is common to hear punters say something like "the odds are 4–5 that horse A will win the race." In probability terms, what we are saying is that the probability of horse A winning the race is 4/5 or 0.80. If we say that "the odds are 4–5 against horse A winning the race" then in probability terms, what we are saying is that the probability of horse A winning the race is 1/5 or 0.20.

11.2.4 Computing Probability: The Addition Rule and the Multiplication Rule

So far, we have been dealing with the computation of the probability of occurrence of single events in isolation. There will be occasions when we are called upon to compute the probability of occurrence of two or more events together. To do this, we need to apply either the addition rule or the multiplication rule.

11.2.5 The Addition Rule

This is a method for finding the probability of occurrence of one event **or** another. More specifically, the addition rule states that

> The probability of occurrence of any one of several possible events is equal to the probability of one event occurring plus the probability of the second event occurring, minus the probability of both events occurring simultaneously.

If there are two possible events A and B, then the general equation for the addition rule is

$$p(A \text{ or } B) = p(A) + p(B) - p(A \text{ and } B)$$

It is important to note that how this equation is applied depends on whether the two events are *mutually exclusive* or not. When two events are mutually exclusive, they cannot occur together. Rolling a die would be an example of a mutually exclusive event. The die cannot land on two sides at the same time; therefore, the probability of each side of the die is mutually exclusive. That is, if you roll a 2 you cannot get any other number on the die. Therefore, if events A and B are mutually exclusive, then the probability of event A and event B occurring together is 0, that is, $p(A \text{ and } B) = 0$. Thus, when two events are mutually exclusive, the general equation for the addition rule can be simplified to

$$p(A \text{ or } B) = p(A) + p(B)$$

Let's do a couple of examples to demonstrate this. In one roll of a die, what is the probability that a 2 or a 5 will turn up? Since the two events (2 and 5) are mutually exclusive (they cannot appear together in one roll of a die), we apply the simplified equation above.

$$p(2 \text{ or } 5) = p(2) + p(5)$$
$$= \frac{1}{6} + \frac{1}{6} = \frac{2}{6}$$
$$= \frac{2}{6} \quad \text{or} \quad 0.3333$$

Thus, the probability of rolling a 2 or a 5 from one roll of a die is 33.33 chances in a hundred.

Let's try another example. If you draw one card from a well-shuffled deck of 52 playing cards, what is the probability that this card is an ace of spade or a 2?

We know that in a deck of 52 playing cards, there is only one ace of spade and 4 cards with the number 2 (2 of spade, 2 of hearts, 2 of diamonds, 2 of clubs). In addition, the two events (ace of spade and 2) are mutually exclusive (they cannot appear together on the same card). Therefore, we apply the simplified equation.

$$p(\text{ace of spade or 2}) = p(\text{ace of spade}) + p(2)$$
$$= \frac{1}{52} + \frac{4}{52} = \frac{5}{52}$$
$$= \frac{5}{52} \quad \text{or} \quad 0.0961$$

Thus, the probability of drawing an ace of spade or a 2 from a deck of 52 playing cards is 9.61 chances in a hundred.

This addition rule for mutually exclusive events can be used for more than two events by simply extending the simplified equation above. For example, if you draw one card from a well-shuffled deck of 52 playing cards, what is the probability that this card is an ace of spade or a 2 or a 9 or a jack of hearts? We know that in a deck of 52 playing cards, there is only one ace of spade, 4 cards with the number 2 (2 of spade, 2 of hearts, 2 of diamonds, 2 of clubs), 4 cards with the number 9 (9 of spade, 9 of hearts, 9 of diamonds, 9 of clubs), and one jack of hearts. In addition, these four events are mutually exclusive (they cannot appear together on the same card). Therefore,

p(ace of spade or 2 or 9 or jack of hearts)

$$= p(\text{ace of spade}) + p(2) + p(9) + p(\text{jack of hearts})$$

$$= \frac{1}{52} + \frac{4}{52} + \frac{4}{52} + \frac{1}{52}$$

$$= \frac{10}{52} \quad \text{or} \quad 0.1923$$

Thus, the probability of drawing an ace of spade or a 2 or a 9 or a jack of hearts from a deck of 52 playing cards is 19.23 chances in a hundred.

When two events are *non-mutually exclusive* they can occur together. In such a situation, we have to apply the full general equation for the addition rule,

$$p(A \text{ or } B) = p(A) + p(B) - p(A \text{ and } B)$$

Let's do a couple of examples to demonstrate this. If you draw one card from a well-shuffled deck of 52 playing cards, what is the probability that this card is a queen or a club? We know that in a deck of 52 playing cards, there are 4 queens (queen of spade, queen of hearts, queen of diamonds, and queen of clubs) and 13 clubs (ace of clubs to king of clubs). This seems to add up to a total of 17 cards. This may suggest that the probability of obtaining a queen or a club is 17/52. However, in arriving at this total we have counted some cards twice. That is, in counting the 4 queens, the count included the queen of clubs. In counting the 13 clubs, the count also included the queen of clubs. In other words, we have counted the queen of clubs twice! This has happened because the two events – queen and club – are not mutually exclusive. They can occur on the same card – the queen of clubs! Obviously, we should count this card only once. Therefore, we should subtract this extra count from the original total, that is, 17 − 1, which gives us the true total of 16 cards. When calculating the probability of occurrence of two events that are non-mutually exclusive, like in this example, we have to apply the full general equation for the addition rule

$$p(A \text{ or } B) = p(A) + p(B) - p(A \text{ and } B)$$

$$p(\text{queen or club}) = p(\text{queen}) + p(\text{club}) - p(\text{queen of clubs})$$

$$= \frac{4}{52} + \frac{13}{52} - \frac{1}{52}$$

$$= \frac{16}{52} \quad \text{or} \quad 0.3076$$

Thus, the probability of drawing a queen or a club from a deck of 52 playing cards is 30.76 chances in a hundred.

Let's try another example. What is the probability of drawing either a spade or an ace from a deck of cards? First, recognize that these two events are not mutually exclusive, that is, they can occur together on the same card – the ace of spade. Therefore, the ace of spade is counted twice when we count the total number of spade cards (13) and the total number of ace cards (4). This has given us the overcounted total of 17 cards. From this total we must subtract the over-counted card of ace of spade, $17 - 1 = 16$ cards. As with the previous example, in calculating the probability of drawing either a spade or an ace from a deck of cards, we have to apply the full general equation for the addition rule

$$p(A \text{ or } B) = p(A) + p(B) - p(A \text{ and } B)$$
$$p(\text{spade or ace}) = p(\text{spade}) + p(\text{ace}) - p(\text{ace of spade})$$
$$= \frac{13}{52} + \frac{4}{52} - \frac{1}{52}$$
$$= \frac{16}{52} \quad \text{or} \quad 0.3076$$

Thus, the probability of drawing a spade or an ace from a deck of 52 playing cards is 30.76 chances in a hundred.

11.2.6 The Multiplication Rule

Sometimes we are faced with the problem of determining the probability of the joint occurrence of two or more events. The multiplication rule is a way to find the probability of two or more events happening at the same time. Note that whereas the addition rule deals with the probability of occurrence of single events (one roll of a die, one draw of a card), the multiplication deals with the probability of the joint occurrence of multiple events (more than one roll or one draw). If there are two events (A and B), then the multiplication rule states that

The probability that events A and B both occur is equal to the probability that event A occurs multiplied by the probability that event B occurs, given that A has occurred.

From the above definition, the general equation for the multiplication rule is expressed as

$$p(A \text{ and } B) = p(A) * p(B/A)$$

For this equation, it is important to note what $p(B/A)$ means. It definitely does not mean the probability of event B divided by event A. What it means is *the probability of occurrence of event B given that event A has occurred.* Thus, $p(B/A)$ is a conditional probability where the probability of event B occurring is conditional upon whether event A has occurred or not. As such, like the addition rule, it is important to note that how this equation is applied depends on the condition of the two events, that is whether the two events are *independent* or *dependent*.

The events are said to be *independent* if the occurrence of one event (A) does not change the probability of occurrence of the event (B) that follows it. Thus, if the occurrence of event A does not affect the probability of occurrence of event B, then

$$p(B/A) = p(B)$$

Therefore, for independent events, the general equation for the multiplication rule can be simplified to

$$p(A \text{ and } B) = p(A) * p(B)$$

Let's do a couple of examples to demonstrate this. A bag contains 10 red marbles and 6 black marbles for a total of 16 marbles. Two marbles are drawn *with replacement* from the bag. What is the probability that both the marbles are black? Let A = the event that the first marble is black; and let B = the event that the second marble is black. Since the first marble (A) drawn is replaced in the bag before the draw of the second marble (B), the two events are independent, that is, the draw of event A does not affect the probability of drawing event B. Therefore, we can apply the simplified equation above

$$p(A \text{ and } B) = p(A) * p(B)$$

$$p(\text{black and black}) = \frac{6}{16} * \frac{6}{16}$$

$$= \frac{36}{256} = 0.1406$$

Thus, the probability of drawing two black marbles with replacement is 14.06 chances in a hundred.

Let's try another example. A dresser drawer contains five neckties of the following colors: blue, brown, red, white, and black. You reach into the drawer and choose a necktie without looking. You replace this necktie and then choose another necktie. What is the probability that you will choose the red necktie both times? Let A = the event that the first necktie selected is red; and let B = the event that the second necktie selected is red. Since the first necktie (A) selected is replaced in the drawer before the second necktie (B) is selected, the two events are independent, that is, the selection of event A does not affect the probability of selecting event B. Therefore, we can apply the simplified equation above

$$p(A \text{ and } B) = p(A) * p(B)$$

$$p(\text{red and red}) = \frac{1}{5} * \frac{1}{5}$$

$$= \frac{1}{25} = 0.04$$

Thus, the probability of selecting two red neckties with replacement is 4 chances in a hundred.

Two events are *dependent* when the occurrence of one event influences the probability of another event. Stated another way, two events are dependent if the occurrence of the first affects the probability of occurrence of the second. In such a case, the general equation for the multiplication rule must be applied, that is,

$$p(A \text{ and } B) = p(A) * p(B/A)$$

Let's do a couple of examples to demonstrate this. A card is chosen at random from a standard deck of 52 playing cards. Without replacing it, a second card is chosen. What is the probability that the first card chosen is a queen and the second card chosen is a jack? Let A = the event that the first card drawn is a queen; and let B = the event that the second card drawn is a jack. Since the first card (A) drawn is not replaced in the deck before the second card (B) is drawn, the two events are dependent, that is, the draw of event A has affected the probability of drawing event B. This is because the draw of the first card (queen) is from the full deck of 52 cards. Since this card is not replaced, there are only 51 cards left in the deck. Thus, the draw of the second card (jack) is from the deck of 51 cards. When the events are dependent, as is in this example, we must apply the full equation for the multiplication rule.

$$p(A \text{ and } B) = p(A) * p(B/A)$$

$$p(\text{queen and jack}) = \frac{4}{52} * \frac{4}{51}$$

$$= \frac{16}{2652} = 0.006$$

Thus, the probability of drawing a queen on the first card and a jack on the second card without replacement is 6 chances in a thousand.

Let's try another example. A bag contains 3 yellow balls and 3 green balls. One ball is drawn from the bag, set aside, and then a second ball is drawn. What is the probability that both balls drawn will be yellow? Let A = the event that the first ball drawn is yellow; and let B = the event that the second ball drawn is yellow. Since the first ball (A) drawn is set aside (not replaced in the bag) before the second ball (B) is drawn, the two events are dependent, that is, the draw of event A has affected the probability of drawing event B. This is because the draw of the first yellow ball is from a total of 6 balls. Since this ball is not replaced, there are only 5 balls left in the bag (2 yellow balls and 3 green balls). Thus, the draw of the second yellow ball is from this total of 5 balls. When the events are dependent, as is in this example, we must apply the full equation for the multiplication rule.

$$p(A \text{ and } B) = p(A) * p(B/A)$$

$$p(\text{yellow and yellow}) = \frac{3}{6} * \frac{2}{5}$$

$$= \frac{6}{30} = 0.20$$

Thus, the probability of drawing a yellow ball on the first draw and a yellow ball on the second draw without replacement is 20 chances in a hundred.

11.2.7 Using the Multiplication and Addition Rules Together

Some probability problems require that we use both the multiplication and addition rules together to solve the problems. For example, suppose I roll two fair dice once, what is the probability that the sum of the numbers showing on the dice will equal 9? There are four possible outcomes that will yield a sum of 9, as follows:

Die1	Die2	Event
6	3	A
3	6	B
5	4	C
4	5	D

First, we calculate the probability of each of the four events (A, B, C, D). Since the roll of the two dice is independent, we apply the simplified multiplication rule.

$$p(A) = p(6 \text{ on die 1 and 3 on die 2})$$
$$p(A) = p(6 \text{ on die 1}) * p(3 \text{ on die 2})$$
$$= \frac{1}{6} * \frac{1}{6} = \frac{1}{36}$$

$$p(B) = p(3 \text{ on die 1 and 6 on die 2})$$
$$p(B) = p(3 \text{ on die 1}) * p(6 \text{ on die 2})$$
$$= \frac{1}{6} * \frac{1}{6} = \frac{1}{36}$$

$$p(C) = p(5 \text{ on die 1 and 4 on die 2})$$
$$p(C) = p(5 \text{ on die 1}) * p(4 \text{ on die 2})$$
$$= \frac{1}{6} * \frac{1}{6} = \frac{1}{36}$$

$$p(D) = p(4 \text{ on die 1 and 5 on die 2})$$
$$p(D) = p(4 \text{ on die 1}) * p(5 \text{ on die 2})$$
$$= \frac{1}{6} * \frac{1}{6} = \frac{1}{36}$$

Any of these four events will yield a sum of 9, $p(\text{sum of } 9) = p(A \text{ or } B \text{ or } C \text{ or } D)$. Moreover, since these events are mutually exclusive, we can use the addition rule for mutually exclusive events to calculate $p(A \text{ or } B \text{ or } C \text{ or } D)$. Thus,

$$p(A \text{ or } B \text{ or } C \text{ or } D) = p(A) + p(B) + p(C) + p(D)$$
$$= \frac{1}{36} + \frac{1}{36} + \frac{1}{36} + \frac{1}{36}$$
$$= \frac{4}{36} = 0.1111$$

Thus, the probability that the sum of the numbers showing on the dice will equal 9 is 11.11 chances in a hundred.

Let's try another example. Suppose you are randomly sampling from three bags of fruits. Bag 1 contains seven different fruits. There is a lemon, a plum, an apple, an orange, a pear, a peach, and a banana. Bags 2 and 3 have the same fruits as Bag 1. If you select one fruit from each bag, what is the probability that you will obtain two apples and a plum? Note that order is not important. All you care about is getting two apples and a plum, in any order.

There are three possible outcomes that will yield 2 apples and a plum, as follows:

Bag 1	Bag 2	Bag 3	Event
Apple	Apple	Plum	A
Apple	Plum	Apple	B
Plum	Apple	Apple	C

First, we calculate the probability of each of the three events (A, B, C). Since these three events are independent, we apply the simplified multiplication rule.

$$p(A) = p(\text{apple in bag 1 and apple in bag 2 and plum in bag 3})$$
$$p(A) = p(\text{apple in bag 1}) * p(\text{apple in bag 2}) * p(\text{plum in bag 3})$$
$$= \frac{1}{7} * \frac{1}{7} * \frac{1}{7} = \frac{1}{343}$$

$$p(B) = p(\text{apple in bag 1 and plum in bag 2 and apple in bag 3})$$
$$p(B) = p(\text{apple in bag 1}) * p(\text{plum in bag 2}) * p(\text{apple in bag 3})$$
$$= \frac{1}{7} * \frac{1}{7} * \frac{1}{7} = \frac{1}{343}$$

$p(C) = p(\text{plum in bag 1 and apple in bag 2 and apple in bag 3})$

$p(C) = p(\text{plum in bag 1}) * p(\text{apple in bag 2}) * p(\text{apple in bag 3})$

$$= \frac{1}{7} * \frac{1}{7} * \frac{1}{7} = \frac{1}{343}$$

Any of these three events will yield 2 apples and a plum, $p(2$ apples and 1 plum$) = p(A$ or B or C$)$. Moreover, since these events are mutually exclusive, we can use the addition rule for mutually exclusive events to calculate $p(A$ or B or C$)$. Thus,

$$p(A \text{ or } B \text{ or } C) = p(A) + p(B) + p(C)$$

$$= \frac{1}{343} + \frac{1}{343} + \frac{1}{343}$$

$$= \frac{3}{343} = 0.00875$$

Thus, the probability of selecting two apples and a plum from the three bags of fruits is 8.75 chances in a thousand.

11.2.8 Computing Probability for Continuous Variables

Up to this point we have considered the calculation of probability for *discrete* variables based on their expected relative frequency. For example, obtaining a 6 on one roll of a die, or drawing a king of spade from a deck of ordinary playing cards. Recall that such probability is defined as

$$p(A) = \frac{\text{Number of events favoring A}}{\text{Total number of events}}$$

However, many of the variables that we deal with in the social sciences are not discrete but *continuous*, for example, IQ, weight, height, test scores. When dealing with continuous variables, it is advisable to express frequency in terms of areas under a normal curve. Thus, for continuous variables, we may express probability as the following equation:

$$p(A) = \frac{\text{Area under the curve corresponding to A}}{\text{Total area under the curve}}$$

Let's do a couple of examples to demonstrate the calculation of probability for continuous variables. Assume that for a general population the mean (μ) IQ score is 100 with a standard deviation (σ) of 16. If you select one person from this population, what is the probability that this person has an IQ score of 132 or higher? (Figure 11.1)

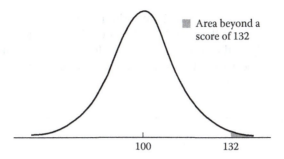

FIGURE 11.1
Proportion of area at or above a score of 132 in a normal distribution with $\mu = 100$ and $\sigma = 16$.

In determining the proportion of area at or above a score of 132, we need to first convert the raw score of 132 to its corresponding standard or z score.

$$z \text{ score} = \frac{X - \mu}{\sigma}$$

$$z \text{ score} = \frac{X - \mu}{\sigma} = \frac{132 - 100}{16} = 2.00$$

Therefore, the person's IQ score of 132 or higher has a corresponding z score value of 2.00. Refer to Table A in the Appendix to determine its corresponding probability. From this z score table, it can be seen that a z score of 2.00 (2 standard deviations above the mean) has a corresponding area of 0.9772 (0.5000 + 0.4772). Therefore, 0.0228 (1 − 0.9772) of the area of the curve lies at or higher than a z score of 2.00. Thus, the probability of randomly selecting a person with an IQ score of 132 or higher is 0.0228 or 2.28 chances in a hundred.

Let's try another example. Assume that we have measured the weight of every person in a certain city and that the mean (μ) weight is 120 pounds and a standard deviation (σ) of 8 pounds. If we randomly select one person from that city, what is the probability that the person's weight will be equal to or less than 106 pounds?

In determining the proportion of area at or below a score of 106, we need to first convert the raw score of 106 to its corresponding standard or z score.

$$z \text{ score} = \frac{X - \mu}{\sigma}$$

$$z \text{ score} = \frac{X - \mu}{\sigma} = \frac{106 - 120}{8} = -1.75$$

Therefore, the person's weight of 106 pounds or less has a corresponding z score value of −1.75. Refer to Table A in the Appendix to determine its corresponding probability. From this z score table, it can be seen that a z score of −1.75 (1.75 standard deviations below the mean) has a corresponding area of 0.4599. Therefore, 0.0401 (0.5000 − 0.4599) of the area lies at or lower than a z score of 1.75. Thus, the probability of randomly selecting a person whose weight is equal to or less than 106 pounds is 0.0401 or 4.01 chances in a hundred.

11.3 Sampling

Why is the technique of sampling important in research? When we conduct research, we do not normally investigate the entire population, unless of course the population is small, like the student population on a university campus. Normally, the research that we conduct is based on very large populations – like the population of Bangkok city, Thailand. To test the entire population of Bangkok residents (~9 million!) will be impossible. As such, the research is normally conducted on a sample drawn from the population of interest. However, at the end of the day, it will be necessary for the researcher to be able to generalize the results obtained from the sample back to the population, that is to say, *what is true of the sample must also be true of the population*. To generalize validly from the sample to the population, the sample cannot be just any subset of the population. Rather, it is critical that the sample is a *random* sample. A random sample is defined as

> A sample in which all members of a group (population or universe) have an equal and independent chance of being selected into the sample.

The sample should be a random sample for two reasons. First, to generalize research findings from a sample to a population, it is necessary to apply the laws of probability to the sample. Unless the sample was generated by a process that assures that every member of the population has an equal chance of being selected into the sample, then we cannot apply the laws of probability (which is crucial for hypothesis testing). Second, once again, in order to generalize from a sample to a population, it is necessary that the sample be representative of the population. One way to achieve representativeness is to choose the sample by a process which assures that all members of the population have an equal and independent chance of being selected into the sample. Some common techniques of random sampling include (1) simple random sampling, (2) stratified proportionate random sampling, and (3) cluster sampling.

11.3.1 Simple Random Sampling

Suppose we have a population of 100 people and we wish to draw a random sample of 20 for an experiment. One way to do this would be to number the individuals in the population from 1 to 100, then write these numbers on 100 slips of paper, put these slips of paper in a barrel, roll it around and then randomly pick out one. Then roll the barrel again and randomly pick out the second slip. Keep on doing this until 20 slips have been randomly picked, and then match the numbers on the slips with the individuals in the population. If the selection of the slips of paper from the barrel is completely random, then we have a random sample of 20 individuals.

11.3.2 Stratified Proportionate Random Sampling

The purpose of stratified proportionate random sampling is to enhance the representativeness of the sample. A simple example will demonstrate this. Suppose the psychology department in your university has a population of 1,000 students stratified/enrolled across four years of study (first year: $n = 500$, second year: $n = 300$, third year: $n = 150$, fourth year: $n = 50$). Let's say you want to draw a representative sample of 100 students. One way to achieve this is to use the simple random sampling technique. From the university registrar, obtain the ID numbers from all 1,000 psychology students, write these numbers on 1,000 slips of paper, put the slips in a barrel, roll it around, and then randomly pick out one slip at a time until you get 100 slips. Match the 100 ID slips with the students and you'll have a random sample of 100 psychology students. However, given that the majority of the students are in first, second, and third year (95% of the population), with very few students in the fourth year (5%), there is a very high probability that your sample will consists mainly of first-, second-, and third-year students, with very few or no fourth-year students. In other words, while you have achieved a random sample, it may not be representative of the student population according to their year of study.

Stratified proportionate random sampling can be used to obtain a sample that is not only random but is also representative of the year of study in the population. Table 11.1 shows how the population of 1,000 students has been stratified across their four years of study. To obtain a stratified proportionate random sample of 100 students:

1. Calculate the proportion of students in each year of study as a proportion of the population of psychology students. Since there are 500 first-year students in the population of 1,000 students,

TABLE 11.1

Stratified Proportionate Random Sampling

Psychology Students ($N = 1,000$)		Proportion	Sample ($n = 100$)
First year:	500	0.50 (500/1000)	$0.50 \times 100 = 50$
Second year:	300	0.30 (300/1000)	$0.30 \times 100 = 30$
Third year:	150	0.15 (150/1000)	$0.15 \times 100 = 15$
Fourth year:	50	0.05 (50/1000)	$0.05 \times 100 = 5$
	$N = 1,000$	1.00	$n = 100$

proportionately these students represent 0.50 (500/1000) of the student population. Similarly, since there are 300 second-year students in the population of 1,000 students, proportionately these students represent 0.30 (300/1000) of the student population. Calculate the proportions for the third- and fourth-year students (0.15 and 0.05, respectively).

2. With these proportions in hand, calculate the number of students in each year of study required for the sample of 100 students. For example, for the 500 first-year students who proportionately represent 0.5 of the student population ($n = 1,000$), they should also proportionately represent 0.5 of the sample ($n = 100$). Thus, the sample should contain 50 first-year students (0.50×100). Similarly, for the 300 second-year students who proportionately represent 0.3 of the student population ($n = 1,000$), they should also proportionately represent 0.3 of the sample ($n = 100$). Thus, the sample should contain 30 second-year students (0.30×100). Calculate the proportionate sample sizes for the third- and fourth-year students in the sample (15 and 5 students, respectively).

3. Once the proportionate sample sizes for all four years of study have been calculated (first year: $n = 50$, second year: $n = 30$, third year: $n = 15$, fourth year: $n = 5$), apply the simple random sampling technique at each level of study to obtain a random and representative (by year of study) sample of 100 students. For example, you want your sample to contain 50 first-year students. From the university registrar, obtain the ID numbers from all 500 first-year psychology students, write these numbers on 500 slips of paper, put the slips in a barrel, roll it around, and then randomly pick out one slip at a time until you get 50 slips. Match the 50 ID slips with the students and you'll have a random sample of 50 first-year psychology students. Repeat this procedure with the second-, third-, and fourth-year students. If at each year of study the selection of students is entirely random, then you will end up with a sample of 100 psychology students that is not only random but also representative in terms of their years of study in the correct proportions within the psychology student population.

11.3.3 Cluster Sampling

The primary purpose of cluster sampling is to lower field costs. For example, let us say there is a city with a population of 20,000 households spread across the entire city, and you want to draw a random sample of 2,000 households. One way to achieve this is to use simple random sampling. From City Hall, obtain the addresses of these 20,000 households, write the addresses on 20,000 slips of paper, put these slips of paper in a barrel, roll it around, and then randomly pick out one slip of paper at a time until you have 2,000 slips. As long as the selection of these addresses is random, you will have a random sample of 2,000 households. However, this technique will spread your sample all across the city resulting in high field costs, time, resources, and manpower. One way to lower field costs is to use cluster sampling to concentrate your sample in specific areas of the population. Suppose that the population of 20,000 households is located in 50 clusters/neighborhoods, with 400 households in each cluster. Therefore, in order to obtain a random sample of 2,000 households, write the names of these 50 clusters on 50 slips of paper, place them in a barrel, roll it around, and then randomly select 5 clusters and interview all 400 households in each cluster ($5 \times 400 = 2,000$). If the selection of the 5 clusters is entirely random, you will end up with a random sample of 2,000 households that are located in specific areas of the city, rather than being spread out all over the city, thus lowering field costs.

11.3.4 Nonrandom Sampling Techniques: Systematic Sampling; Quota Sampling

It is also important to note that some popular sampling techniques are nonrandom and therefore the resultant sample is nonrepresentative. Examples of nonrandom sampling techniques include *systematic sampling* and *quota sampling*.

Systematic sampling—Examples of systematic sampling include techniques such as standing in front of the university library and interviewing every 10th student who exits the library; or in a telephone survey, call up every 5th person in the telephone book. These techniques are nonrandom because not everyone in the population has an equal chance of being selected into the sample.

Quota sampling—This technique is sometimes called 'the poor man's stratified sampling'. With this technique, the researcher specifies specific characteristics of the sample that he/she wants. For example, the researcher may specify that he wants in his sample only those participants who possess the following characteristics: male, 28 years old, Catholic, and with a master's degree in Economics. While the sample may satisfy the requirements of the researcher, it is definitely nonrandom.

11.3.5 Sampling with or without Replacement

Suppose we have 10 balls numbered from 1 to 10 inside a barrel. We want to select a random sample of three balls from the barrel. After we pick the first ball from the barrel, we can put the ball aside or we can put it back into the barrel. If we put the ball back in the barrel, it may be selected more than once; if we put it aside, it can be selected only one time. When a population element can be selected more than one time, we are *sampling with replacement*. When a population element can be selected only one time, we are *sampling without replacement*.

Sampling with or without replacement has important implications for the calculation of probability. For the above example, say we want to calculate the probability of selecting the three balls numbered 1, 5, and 9 with replacement. As each selected ball is replaced before the draw of the next ball, the three events are said to be independent because the draw of the first ball does not affect the probability of occurrence of the event that follows it. Thus, the multiplication rule for independent events can be applied.

$$p(A \text{ and } B \text{ and } C) = p(A) * p(B) * p(C)$$

$$p(1 \text{ and } 5 \text{ and } 9) = \frac{1}{10} * \frac{1}{10} * \frac{1}{10} = \frac{1}{1000}$$

$$= 0.001$$

Thus, the probability of selecting ball numbered 1 on the first draw and ball numbered 5 on the second draw and ball numbered 9 on the third draw with replacement is 1 chance in a thousand.

The above example is in contrast to the next and similar example but one that deals with sampling without replacement. Suppose we want to calculate the probability of selecting the three balls numbered 1, 5, and 9 without replacement. As each selected ball is set aside before the draw of the next ball, the three events are said to be dependent because the draw of the first ball affects the probability of occurrence of the event that follows it. Thus, the multiplication rule for dependent events can be applied.

$$p(A \text{ and } B \text{ and } C) = p(A) * p(B/A) * p(C/AB)$$

$$p(1 \text{ and } 5 \text{ and } 9) = \frac{1}{10} * \frac{1}{9} * \frac{1}{8} = \frac{1}{720}$$

$$= 0.0014$$

Thus, the probability of selecting ball numbered 1 on the first draw and ball numbered 5 on the second draw and ball numbered 9 on the third draw without replacement is 1.4 chances in a thousand.

11.4 Confidence Interval and Confidence Level

As mentioned at the beginning of this chapter, a major aim of conducting research is to draw inferences/conclusions about a population based on information obtained from a sample taken from that population. For example, the purpose of taking a random sample from a population and computing a statistic, such as the mean from the data, is to estimate the mean of the population. As such, a primary purpose of inferential statistics is to enable the researcher to infer from the sample data some characteristics of the population. However, how well the sample statistic estimates the underlying population value is always an issue. Let's take an example to demonstrate this problem.

Say you were interested in the mean weight of 10-year-old boys living in Thailand. Since it would have been impractical to weigh all the 10-year-old boys in Thailand, you took a sample of 500 and found that the mean weight was 45 kg. This sample mean of 45 is a *point estimate* of the population mean (*a point estimate of a population parameter is a single value used to estimate the population parameter, e.g., the sample mean $\bar{x}$ is a point estimate of the population mean*). While the point estimate of 45 kg is true for the sample of 10-year-old boys, the estimate by itself is of limited usefulness because of the uncertainty associated with the estimate; you simply do not know how far this sample mean deviates from the population mean. In other words, how confident are you that the sample mean represents the true population mean 'plus or minus' say 2 kg from 45? You simply do not know.

Whenever we use sample statistics to infer a population parameter, the accuracy of the sample statistics in estimating the underlying population value is always an issue. In addressing this accuracy issue, we employ the *confidence interval* that provides a range of values, which is likely to contain the population parameter of interest, given a specified *confidence level*.

The *confidence interval* (also called the margin of error), which is calculated from sample data, gives an estimated range of values which is likely to include an unknown population parameter. It is the + or − figure usually reported in newspaper or television opinion poll results. In the above example, say you computed a confidence interval of ±2 kg. If you find that the mean weight of your sample is 45 kg, then the true estimated mean weight of the entire population of 10-year-old boys in Thailand will lie between 43 kg (45 − 2) and 47 kg (45 + 2).

The *confidence level* for a confidence interval specifies the probability that the confidence interval produced will contain the true population value. The confidence level tells you how sure you can be. It is expressed as a percentage and represents how confident the researcher is that the population parameter (e.g., mean weight) lies within the confidence interval. Confidence levels are commonly specified as 90% (0.90), 95% (0.95), and 99% (0.99). The 90% confidence level means you can be 90% certain; the 95% confidence level means you can be 95% certain; the 99% confidence level means you can be 99% certain. Let's say we have calculated the 95% confidence interval for the example above to be $43 < \mu < 47$. In effect, what we are saying is that we are 95% confident that the population mean weight lies within this interval, that is, between the two bounds of 43 and 47 kg. Indeed, if we were to draw repeated samples and the 95% confidence interval calculated for each sample, then in all likelihood, 95% of the computed intervals will contain the population mean. The converse is equally true. That is, 5% of the intervals will not contain the population mean.

11.4.1 How to Calculate the Confidence Interval

Suppose we are interested in finding the overall population mean weight of all students at a university. Since we do not want to measure the weights of the entire student population, we choose a random sample 100 students and find that the mean weight ($\bar{x}$) is 61.459 kg, with a standard deviation of 14.2. What is the 95% confidence interval for the population mean? That is, $? < \mu < ?$ The following presents a step-by-step solution to this problem.

1. Let $z(a/2)$ be the standard (z) score for the confidence level (95% or 0.95) for a two-tailed distribution (the two- and one-tailed distribution is explained in Chapter 12).

   $\bar{x}$ is the sample mean = 61.459

   s is the standard deviation = 14.2

   n is the sample size = 100

2. Calculate the margin of error: $E = z(a/2) * {}^s\!\sqrt{n}$

   We know that $s = 14.2$ and $n = 100$, but we need to find $z(a/2)$.

3. To find $z(a/2)$, we need to first divide the confidence level of 0.95 by 2 ($a/2$) to get 0.475 (two-tailed distribution). Next, we look at the z table (Table A in the Appendix) to find the corresponding z value that goes with 0.475. You'll see that the closest value is 1.96.

4. Now that we know $z(a/2) = 1.96$, we can go back to the formula in Step 2 to calculate the margin of error. The margin of error is $E = z(a/2) * {}^s\!\sqrt{n}$

$$\text{Thus, } E = 1.96 * {}^{14.2}\!\sqrt{100}$$
$$= 2.7832$$

5. Finally, the 95% confidence interval is

   $\bar{x} \pm E = 61.459 \pm 2.7832$

   $= 61.459 + 2.7832 = \textbf{64.2422}$ (upper confidence interval limit)

   and

   $= 61.459 - 2.7832 = \textbf{58.6758}$ (lower confidence interval limit)

That is, we are 95% confident that the interval 58.6758 and 64.2422 contains the true population weight mean.

11.4.2 SPSS Windows Method

1. Launch the SPSS program and then open the data file **EX14.SAV** (this file contains the 100 weight scores). From the menu bar, click

Analyze, and then **Descriptive Statistics**, and then **Explore**... The following **Explore** Window will open.

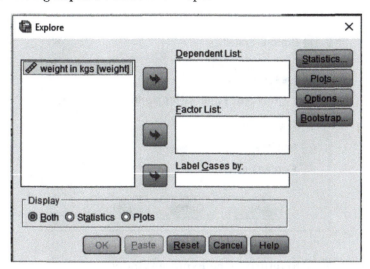

2. In the left-hand field containing the study's **WEIGHT** variable, click (highlight) this variable, and then click ⬇ to transfer the selected **WEIGHT** variable to the **Dependent List:** field.

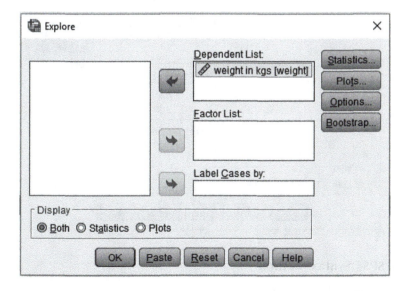

3. Click [Statistics...] to open the **Explore Statistics** dialogue box. Check the **Descriptives** cell. Type 95 in the **Confidence interval for mean**

cell to obtain the 95% confidence interval for the estimated population mean.

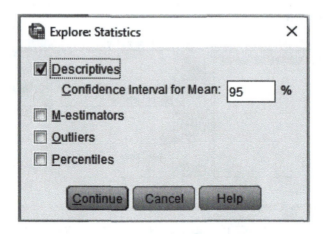

Click [Continue] to return to the **Explore** Window.

4. Click [OK] to run the analysis. See Table 11.2 for the results.

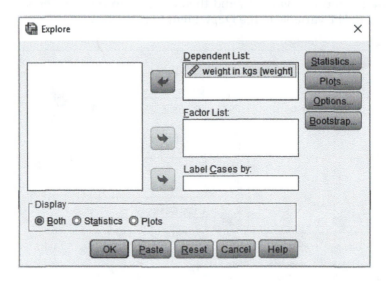

11.4.3 SPSS Syntax Method

```
EXAMINE VARIABLES=WEIGHT
/STATISTICS DESCRIPTIVES
/CINTERVAL 95
/MISSING LISTWISE
/NOTOTAL.
```

1. From the menu bar, click **File**, then **New**, and then **Syntax.** The following **IBM SPSS Statistics Syntax Editor** Window will open.

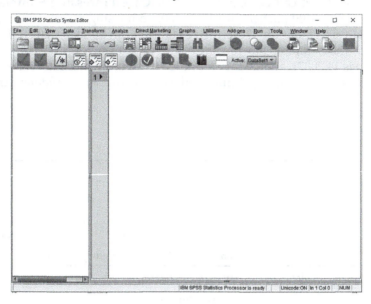

2. Type the **Examine** analysis syntax command in the **IBM SPSS Statistics Syntax Editor** Window.

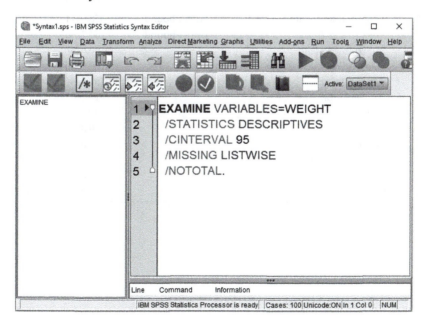

3. To run the **Examine** analysis, click ▶ or click Run and then **All**.

11.4.4 SPSS Output

From Table 11.2, it can be seen that under **Case Processing Summary** the entire sample of 100 participants ($N = 100$, no missing cases) was employed in the analysis. Under the **Descriptives** table, it can be seen that (1) the mean weight of the sample's 100 10-year-old boys is 61.459 kg, (2) the standard deviation is 14.20021, and (3) the computed 95% confidence interval is 58.6414 (lower confidence interval limit) and 64.2766 (upper confidence interval limit).

TABLE 11.2

Explore Output

Explore							
Case Processing Summary							
				Cases			
		Valid		**Missing**		**Total**	
		N	**Percent**	*N*	**Percent**	*N*	**Percent**
Weight in kg		100	100.0	0	0.0	100	100.0

				Statistic	**Std. Error**
Descriptives					
Weight in kg	Mean			61.4590	1.42002
	95% Confidence interval for mean		Lower bound	58.6414	
			Upper bound	64.2766	
	5% trimmed mean			60.6767	
	Median			60.0000	
	Variance			201.646	
	Std. deviation			14.20021	
	Minimum			37.00	
	Maximum			120.00	
	Range			83.00	
	Interquartile range			19.50	
	Skewness			1.001	0.241
	Kurtosis			1.974	0.478

12

Introduction to Hypothesis Testing

12.1 Introduction to Hypothesis Testing

Fundamental to the strategy of science is the formulation and testing of hypotheses about populations or the effects of experimental conditions on criterion variables. At the core of the scientific methodology is an experiment, and part and parcel of conducting an experiment is to test the predictions/hypotheses generated from a particular theory or past research findings or from the researcher's literature review. In testing a hypothesis, data must be collected and then analyzed. For example, a researcher may be interested in testing the hypothesis that there is gender difference in problem-solving ability. To test this hypothesis, he employs a simple *between-groups design* comprising two groups of 5 male subjects and 5 female subjects. Table 12.1 presents each subject's problem-solving score and the two groups' average scores.

As can be seen, the male subjects scored on average 12 points whereas the female subjects scored on average 88 points, an average difference of 76 points. Sometimes the result (such as this) is so clear-cut that we do not need statistical analysis to inform us that there is gender difference in problem-solving ability, and that females are better problem-solvers than males. However, clear-cut results such as this are rare, and more often than not we will obtain results that are much closer, as indicated in Table 12.2.

Now, we find that the average problem-solving score for the male subjects is **85** and for the female subjects, it is **90**, a difference of only **5** points. Based on these results, can we still conclude that there is gender difference in problem-solving ability? Indeed, can we conclude that females are better problem-solvers than males? As you can see, the answer is not so clear-cut now. That is, although the results clearly showed that there is a difference between the sample means with females scoring higher on average than males it is difficult to draw a definitive conclusion about gender difference on the basis of a mean score difference of just 5 points. The reason why it is difficult to make such a decision is the possibility that the observed difference could have been due to the *chance variability* of problem-solving skills among males and females. Take a look at the problem-solving scores in Tables 12.1 and

TABLE 12.1

Problem-Solving Scores as a Function of Gender

	Male		Female
s1	10	s1	85
s2	5	s2	90
s3	15	s3	95
s4	10	s4	90
s5	20	s5	80
Average scores	**12**		**88**
(An average difference of **76** points)			

12.2. Even *within* each group of males and females, there are differences in their problem-solving scores. Also, *between* the two groups of males and females, regardless of their gender, there are also differences, because people are different in how they respond to problems to be solved. These are naturally occurring (or what we call *chance*) differences that have nothing to do with the subject's gender (or other characteristics). So how do we determine whether the obtained result is reliable? Of course, if we repeat this study a hundred times and every time we obtain the same or similar result, that is females scored higher than males, then we can be reasonably confident about the reliability of our finding. However, repeating experiments over and over again to test the reliability of the obtained results is not a practical solution. As researchers, we are often constrained by time, resources, finances, and manpower. Therefore, a more efficient way to determine whether the result that we have obtained is due to *chance-variation* or to the effect of the IV (*gender*) is to employ inferential statistics (based on probability theory) to help us make a decision about the outcome. The critical questions that must be answered by inferential statistics then become *"does the obtained difference represent a reliable and meaningful difference, or is it due purely to chance variation and therefore without consistency?"* A prime function of hypothesis testing via inferential statistics is to provide rigorous and logically sound procedures for answering these questions.

TABLE 12.2

Problem-Solving Scores as a Function of Gender

	Male		Female
s1	80	s1	90
s2	75	s2	90
s3	90	s3	95
s4	85	s4	90
s5	95	s5	85
Average scores	**85**		**90**
(An average difference of **5** points)			

12.2 Types of Hypotheses

Before discussing the procedures of hypothesis testing, it is necessary to distinguish between the *research/alternative hypothesis* and the *null hypothesis*.

12.2.1 Research/Alternative Hypothesis

Hypotheses derived from the researcher's theory about some social phenomenon are called research or alternative hypotheses. For the example above, the research hypothesis would be stated as *"there is gender difference in problem-solving ability."* The researcher usually believes that his research hypotheses are true, or that they are accurate statements about the condition of things he is investigating. He believes that his hypotheses are true to the extent that the theory from which they were derived is adequate. However, theories are only suppositions about the true nature of things, and thus hypotheses derived from theories must also be regarded as just tentative suppositions about things until they have been tested. Testing hypotheses means subjecting them to *confirmation* or *disconfirmation*.

12.2.2 Null Hypothesis

Null hypotheses are, in a sense, the reverse of research hypotheses. They are also statements about the reality of things, except that they serve to *deny* what is explicitly indicated in a given research hypothesis. For example, if the researcher states as his research hypothesis that *"there is gender difference in problem-solving ability,"* he may also state a null hypothesis that can be used to evaluate the accuracy of his research hypothesis. The null hypothesis would be *"there is no gender difference in problem-solving ability, that is, the obtained gender difference is a chance outcome."* If the researcher is able to demonstrate that the probability of the mean difference problem-solving score occurring by chance alone is very small, then he concludes that the null hypothesis is refuted or regarded as not true. If the researcher rejects the null (chance) hypothesis, then logically the statement that *"there is gender difference in problem-solving ability"* is supported. In other words, the researcher constructs a situation that contains two contradictory statements, namely *"there is/is no gender difference in problem-solving ability."* These statements are worded in such a way that they are mutually exclusive, such that the confirmation of one is the denial or refutation of the other. *Both cannot coexist simultaneously.* Thus, if we reject the null hypothesis, then we must accept the research hypothesis as being supported. If we fail to reject the null hypothesis, then we have to reject the research hypothesis and conclude that the obtained result is probably a chance outcome. It is worthwhile noting that in social research, when we test a hypothesis, we always test the null hypothesis – the statement that there is no difference between groups or no relationship between variables (i.e., the obtained difference or relationship is a chance outcome). Whether our research hypothesis is supported or refuted depends on the outcome of the test of the null hypothesis.

Why do we test the null hypothesis and not the research hypothesis? We do not test the research hypothesis because the research hypothesis is essentially a statement of 'fact' about a phenomenon that the researcher believes to be true, that is, 'there is gender difference in problem-solving ability'. There is no statistics associated with statements of fact. The null hypothesis on the other hand is a 'chance' hypothesis, which states that any observed gender difference is due to chance only, and we can always calculate the probability of chance occurrences.

12.2.3 Hypotheses: Nondirectional or Directional

In specifying a research hypothesis, we can specify either a nondirectional or a directional hypothesis. A *nondirectional research hypothesis* simply predicts that there will be a difference between, say, two groups but does not specify how the groups will differ. Using the problem-solving ability example above, the researcher can specify a nondirectional hypothesis stated as *"there is gender difference in problem-solving scores."* While this hypothesis predicts that there will be gender difference in subjects' problem-solving scores, it does not specify the direction of the difference, that is, whether males will score higher or females will score higher.

A *directional hypothesis* not only predicts that there will be a difference between the two groups but also it specifies how the two groups will differ, that is, the direction of the difference. It employs comparison/relative terms such as "greater than," "less than," "stronger than," or "weaker than." Once again, using the problem-solving ability example above, the researcher can specify a directional hypothesis stated as *"females will score higher than males in their problem-solving scores."* This hypothesis predicts that not only will there be gender difference in problem-solving scores, but also that females will score higher.

12.3 Testing Hypotheses

Testing hypotheses means subjecting them to some sort of empirical scrutiny to determine whether they are supported or refuted by what the researcher observes. Take for example, the problem-solving ability experiment above hypothesizes that there is gender difference in problem-solving scores (nondirectional research hypothesis). The obtained data indicated that the average problem-solving score for the male subjects is 85 and for the female subjects, it is 90, a difference of 5 points (see Table 12.2). The question facing the researcher then is, what are the bases upon which he concludes that his hypothesis is supported or refuted? Given this finding, isn't it possible that two researchers working independently may arrive at quite different conclusions pertaining to the obtained finding? The answer is a definite 'yes!' Therefore, in arriving at a conclusion pertaining to a research finding, it is necessary to avoid or at least reduce the amount of subjectivity that exists when research findings are interpreted. Social scientists do this by employing a decision rule that specifies the conditions under which researchers will decide to refute or support the hypotheses they are testing. This decision rule is called the *level of significance.*

12.3.1 Level of Significance

In our problem-solving ability example, we noted that the average problem-solving score for the male subjects is 85 and for the female subjects, it is 90, a difference of 5 points. When a difference in characteristics (in this example, problem-solving scores) between two groups is observed, at what point do we conclude that the difference is a *significant* one (i.e., not due to chance)? We are usually going to observe differences between people regarding commonly held characteristics, but how do we decide that the differences mean anything important to us? That is, how do we judge these differences? What may be a significant difference to one researcher may not be considered as such by another researcher. In order that we may introduce greater objectivity into our interpretations of observations, we establish a standard against which we judge our research findings. This standard is called the *level of significance* or the *alpha (α) level*. To state a level of significance (α level) is to state a probability level at which the researcher will decide to accept or reject the null hypothesis. How do levels of significance operate to enable researchers to make decisions about their observations? To answer this question, we need to look at probability theory.

As mentioned Chapter 11, probability in social research is concerned with the possible outcomes of experiments, that is, the likelihood that one's observations or results are expected or unexpected. In understanding how we use probability to help us make a decision regarding the test of our hypothesis, it is important to note once again that ***when we test a hypothesis, we always test the null (chance) hypothesis, and any conclusion we draw about the research hypothesis is based on the outcome of the test of the null hypothesis.*** To recap, in our problem-solving ability example, the nondirectional research and null hypotheses are stated as follows:

- *Research/alternative hypothesis (Ha).* There is gender difference in problem-solving behavior.
- *Null hypothesis (Ho).* There is no gender difference in problem-solving ability, that is, any observed difference is due to chance.

(Recall that the average problem-solving score for males is 85 and for females it is 90, a difference of 5 points.)

Suppose that the probability of getting the result that we have obtained (i.e., male: average score = 85 and female: average score = 90, a difference of 5 points) being due to chance alone is 1 time in a million (0.000001). Since this chance occurrence is so small, we would no doubt reject chance as an explanation of the observed gender difference in the average scores. And once we reject chance (the null hypothesis) as an explanation for the result, we have to accept the research hypothesis that gender did indeed have an effect on the problem-solving scores. That is, based on the rejection of the null or chance hypothesis, we have to accept the research hypothesis, because it is the only other possible explanation.

However, if I were to say to you that the probability of getting the result that we have obtained (i.e., male: average score = 85 and female: average score = 90, a difference of 5 points) being due to chance alone is not 1 time in a million but 1 in 4 (0.25), can we still reject chance as a cause of our result? Given the probability that the difference in the average scores for males and females being due to chance is quite high (0.25 or 1 in 4), the decision is not so clear-cut this time. Indeed, with such a high chance outcome we will not be able to reject the null hypothesis of 'chance difference'. Failure to reject the null hypothesis, that is retaining the null hypothesis means that we have to reject the research hypothesis and conclude that the observed difference in problem-solving scores between males and females is due to chance, that is, there is no significant gender difference.

While we employ probability theory to assist us in making a decision as to whether to reject or to retain the null hypothesis, there is still the question of at what chance probability level (α) should we reject the null hypothesis? For example, if the calculated probability of the 5-point difference between males and females in our problem-solving ability example occurring by chance alone is 10 times in a hundred (0.10), can we reject chance (the null hypothesis) as an explanation? What if the probability of the 5-point difference occurring by chance is 5 times out of a hundred (0.05), can we reject chance as an explanation? Indeed, what will prevent two researchers faced with the same research outcome from drawing very different conclusions about their hypotheses if they use different chance probability levels to evaluate their null hypotheses? In order to introduce some objectivity in the decision-making process, conventional decision rules have been set up to assist researchers in deciding when to reject the null hypothesis. These decision rules relate to the levels of significance (probability of chance occurrence) that are used to evaluate the null hypothesis. There are three conventional α levels of significance: $p \leq 0.05$; $p \leq 0.01$; $p \leq 0.001$.

Using $p \leq 0.05$ as an example, what this level of significance says is that 'if the probability of obtaining the result that we have obtained (in our example, a difference of 5 points in problem-solving scores between males and females) by chance alone is equal to or less than 5 times out of hundred, then we can reject chance as an explanation. Once we reject chance (the null hypothesis) as an explanation, we have to accept the research hypothesis'.

Applying the above $p \leq 0.05$ decision rule to the results from our example:

- Step 1 – Calculate the probability of getting the result obtained (male: average score = 85, female: average score = 90) if chance alone is responsible. Please note that the calculation differs depending on the statistical test used.

- Step 2 – Compare this chance probability value against the $p \leq 0.05$ α level of significance. For example, if the probability of getting the above result by chance alone is 0.02 (2 times in a hundred), and since this chance probability occurrence is smaller than the 0.05 criterion, we can reject the null (chance) hypothesis. Once we reject the null hypothesis we have to accept the research hypothesis, and thus conclude that gender has an effect on problem-solving ability. On the other hand, if the probability of getting the above result by chance alone is 0.17 (17 times in a hundred), and since this chance probability occurrence is bigger than the 0.05 α criterion, we have to accept the null (chance) hypothesis. Once we accept the null hypothesis to be true, we have to reject the research hypothesis, and thus conclude that gender has no effect on problem-solving ability (the difference of 5 points between males and females is due to chance).

12.3.2 Two-Tailed and One-Tailed Test of Significance

Testing nondirectional and directional hypotheses is associated with the use of the *two-tailed* and *one-tailed* test of significance. A two-tailed test (nondirectional test) will test both sides of the probability curve. Using the earlier example of gender difference in problem-solving scores, we note that the mean problem-solving score for males is 85 and that for females is 90 (see Table 12.2). In order to test whether these mean scores are significantly different from each other (i.e., not due to chance), the researcher may set the probability level (α) for rejecting the null (chance) hypothesis at $p \leq 0.05$. If the research hypothesis merely specifies that there is gender difference in problem-solving ability without specifying the direction of the difference (i.e., whether males will score higher or females will score higher), then a two-tailed test of significance, which will test both sides of the probability curve, is appropriate. This requires that the set probability value of $p \leq 0.05$ be divided equally between the two sides of the curve ($p \leq 0.025$) as indicated in Figure 12.1.

Thus, for a two-tailed test of significance, the researcher can conclude that the mean problem-solving scores for males (85) and for females (90) are significantly different from each other (not due to chance) if the test statistics (e.g., z-score) is in the top 2.5% or bottom 2.5% of its probability distribution, resulting in a p-value less than 0.05.

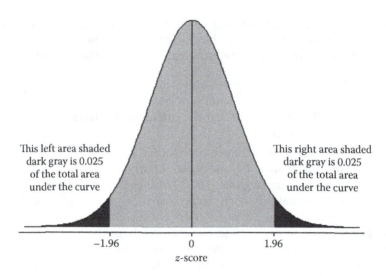

This left area shaded
dark gray is 0.025
of the total area
under the curve

This right area shaded
dark gray is 0.025
of the total area
under the curve

−1.96   0   1.96

z-score

FIGURE 12.1
Two-tailed test of significance.

A one-tailed test (directional test) will test only one side of the probability curve. Once again, using the example of gender difference in problem-solving scores, the research hypothesis may specify not only gender difference in the problem-solving scores but also the direction of the difference. For instance, if the research hypothesis states that females will score higher than males in their problem-solving scores, then the appropriate test is the one-tailed test of significance which will test only one side of the probability curve. Thus, if the researcher sets the probability level (α) for rejecting the null (chance) hypothesis at $p \leq 0.05$, the one-tailed test will test (1) whether the mean problem-solving score for females (90) is significantly higher than the mean problem-solving score for males (85), or (2) whether the mean problem-solving score for males (85) is significantly lower than the mean problem-solving score for females (90), but not both (see Figure 12.2).

Thus, for a one-tailed test of significance, the researcher can conclude that the mean problem-solving score for females (90) is significantly higher than that for males (85) if the test statistics (e.g., z-score) is in the top 5% of its probability distribution, resulting in a p-value less than 0.05.

12.3.3 Type I and Type II Errors

As stated earlier, hypothesis testing involves the use of the level of significance (α) to test a null hypothesis. The null hypothesis is either true or false and represents the default claim for whether an experimental treatment has an effect or not. For example, when examining the effectiveness of a drug, the null hypothesis would be that the drug has no effect on a disease. However, no hypothesis test is 100% certain. Because hypothesis testing is based on

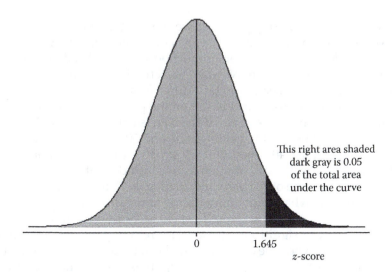

This right area shaded dark gray is 0.05 of the total area under the curve

0   1.645

z-score

FIGURE 12.2
One-tailed test of significance.

probabilities – testing the null or chance hypothesis and not the research hypothesis – there is always a chance of drawing an incorrect conclusion. In an ideal world, we would always reject the null hypothesis when it is false, and we would not reject the null hypothesis when it is indeed true. But there are two other scenarios that are possible, each of which will result in an error.

Type I error. *If we reject the null hypothesis when it is true and should not be rejected, we have committed Type I error.* Let's use a coin example to demonstrate this. Say we want to test whether a particular coin is biased or neutral. A coin has only two sides – head and tail. If the coin is neutral and if we flip the coin once, then the probability of getting a head is the same as the probability of getting a tail, that is, $p(H) = 0.5$ and $p(T) = 0.5$. If the coin is biased, then in all likelihood the probability of getting a head is not the same as the probability of getting a tail, that is, $p(H) \neq 0.5$ and $p(T) \neq 0.5$. These probabilities can be stated as the research and null hypotheses. Thus,

- Research hypothesis – the coin is biased, that is, $p(H) \neq p(T) \neq 0.5$.
- Null hypothesis – the coin is neutral, that is, $p(H) = p(T) = 0.5$.

Say for this example we set the level of significance (α) for rejecting the null hypothesis at $p \leq 0.05$. We are now ready to start the test of our hypothesis, that is, whether our coin is biased or neutral.

Recall that hypothesis testing involves testing the null hypothesis, which at the start of the experiment is always assumed to be true. Thus, if the null hypothesis is true, that is if the coin is neutral, this assumption of neutrality is reflected in the expected probabilities of getting a head or a tail in one flip of the coin, that is, $p(H) = p(T) = 0.5$. Therefore, if we were to flip the coin 100 times, we would expect that we would get 50 heads and 50 tails. This expected outcome assumes that the coin is not biased. A biased coin, however, would reflect a greater proportion of either heads or tails compared with an unbiased one. Now, if we actually flip the coin 100 times, does the distribution of heads and tails differ from what would be expected? Determining what is expected is based on the possible outcomes associated with our observations. Once again, if the coin is neutral and we flip the coin 100 times we would expect to get 50 heads and 50 tails. What if it comes up head 60 times and tail 40 times, will we say that this distribution is so different from our expected distribution as to conclude that the coin is biased? How about an outcome of 70 heads and 30 tails, or 75 heads and 25 tails? The question here is, at what point do we decide to regard an outcome as significantly different from what we would expect according to probability? If we set the probability or significance (α) level for rejecting the null hypothesis at $p \leq 0.05$, then only when the coin comes up head (or tail) 95 times or more out of 100 flips, can we then reject the null hypothesis of neutrality and accept the research hypothesis that the coin is biased. In statistical terms, we say that the probability of the observed outcome of 95 heads (or 95 tails) from 100 flips occurring by chance is equal to or less than five times out of a hundred, that is, we conclude that something else other than chance has affected the outcome – the coin is biased.

In this coin example, we use the level of significance to help us decide whether to accept or reject the null hypothesis. That is, when we set the level of significance at 0.05, we will only reject the null hypothesis of neutrality if the coin turns up heads 95 times or more (or tails 5 times or less) in 100 flips. However, if the coin turns up heads 96 times out of 100 flips, and we reject the null hypothesis of neutrality, isn't it possible that we could be wrong in rejecting the null hypothesis? Is it not possible that by some incredible stroke of luck we get a run of 96 heads out of 100 flips? Highly unlikely but definitely not impossible! The answer to this question therefore must be "yes."

If we reject the null hypothesis of neutrality when it is true and should not be rejected, we have committed Type I error. In testing a hypothesis, the level of significance (α) set to decide whether to accept or reject the null hypothesis is the amount of Type I error the researcher is willing to permit. When we employ the 0.05 level of significance, approximately 5% of the time we will be wrong when we reject the null hypothesis and assert its alternative. It would seem then, that in order to reduce this type of error, we should set the rejection level as low as possible. For example, if we were to set the level of significance at 0.001, we would risk a Type I error only about one time in every thousand. However, the lower we set the level of significance, the greater is the likelihood that we will make a Type II error.

Type II error. *If we fail to reject the null hypothesis when it is actually false, we have committed Type II error.* This type of error is far more common than a Type I error. For example, in the coin experiment, if we set the level of significance at 0.01 for rejecting the null hypothesis, and if the coin turns up heads

TABLE 12.3

Relationship between Type I and Type II Errors

Researcher's Decision	Null Hypothesis (Ho) is True	Null Hypothesis (Ho) is False
Accept Ho	Correct decision	**Type II error**
Reject Ho	**Type I error**	Correct decision

98 times out of 100 flips, the researcher will not be able to reject the null hypothesis of neutrality. That is, based on these observations, the researcher cannot claim that the coin is biased (the alternative hypothesis) even though it may very well be. Only when the coin turns up heads 99 times or more will the researcher be able to reject the null hypothesis. However, in all likelihood, if the coin turns up heads 97 times or 98 times in 100 flips, the coin is indeed biased. But the researcher is still forced to accept the null hypothesis of neutrality because of the 0.01 level of significance that was set (i.e., only when the coin turns up heads 99 times or more will the researcher be able to reject the null hypothesis). By accepting the null hypothesis of neutrality before the coin turns up heads 99 times or more, the researcher has, in all likelihood, committed Type II error.

It is clear then, that Type I and Type II errors cannot be eliminated. They can be minimized, but minimizing one type of error will increase the probability of committing the other error. The lower we set the level of significance, the less is the likelihood of a Type I error, and the greater is the likelihood of a Type II error. Conversely, the higher we set the level of significance, the greater the likelihood of a Type I error, and the smaller the likelihood of a Type II error. The relationship between Type I and Type II errors can be demonstrated in Table 12.3.

13

Hypothesis Testing: t test for Independent and Correlated Groups

13.1 Introduction to the *t* test

In conducting research, we are often faced with the question of whether two groups of participants in a sample differ significantly on a measurement variable. That is, whether the observed difference represents a real/meaningful difference or whether the difference is merely a chance outcome. For example,

> Do males and females differ in performance on a standardized achievement test?
>
> What is the effect of drug versus no drug on rats' maze learning behavior?
>
> Does the recidivism rate of juvenile offenders who are provided with father-figures differ from those without father-figures?

These research questions can be answered by comparing the mean scores from the two groups of participants. This involves using inferential statistical tests to test whether the mean score difference represents a real (i.e., statistically significant) difference or whether the obtained difference is the result of chance variation between the groups. The analysis that is most appropriate whenever we want to compare the means of two independent or correlated groups is the *t* test. The question the *t* test addresses is whether the means are statistically different.

But what does it mean when we say that the mean scores for two groups are statistically different? Take a look at Figure 13.1 that depicts three situations in which the difference between the means is the same in all three. However, the shape of the distributions is quite different due to the differences in their variability. The top example shows two distributions with moderate variability of scores within each group. The second situation shows two distributions with high score variability within the two groups. The third situation shows two distributions with low score variability within the two groups.

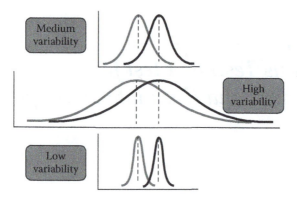

FIGURE 13.1
Differences between group means relative to their variability.

In assessing which of the three situations is most likely to yield a significant group difference in mean scores it is apparent that the low-variability situation is the case most likely to yield a significant difference. Why? Because compared to the moderate- and high-variability cases, there is relatively little overlap between the two bell-shaped curves in the low-variability case. It can be seen that in the high-variability case, there is very little group difference because the two bell-shaped distributions overlap so much.

The above scenarios demonstrate a very important point when comparing group means. In assessing the mean score difference for two groups, we have to assess the difference between their means relative to the spread or variability of their scores. Indeed, the relationship between group means and their variability is reflected in the formula for the t test. The formula for the t test is essentially a ratio in which the top part of the ratio is just the difference between the two groups' mean scores. The bottom part is a measure of the variability or dispersion of the scores.

Another important point to note is that there are actually two types of t test and their employment depends on the design of the study and the associated research questions/hypotheses posed. These are the *independent t test* and the *dependent/correlated t test*.

The *independent t test* is appropriate when the research requires the comparison of the means between two unrelated groups on the same continuous, dependent variable. For example, you could use an independent t test to test whether the IQ of 12-year-old children in Thailand differed based on gender. In this scenario, the dependent variable would be "IQ scores" and the independent variable would be "gender" (male/female). Alternately, you could use the independent t test to test whether there is a significant difference in stress level based on educational level. For this scenario, the dependent variable would be "stress level" and the independent variable would be "educational level" (undergraduates/postgraduates).

The *dependent/correlated t test* compares the means of two 'related groups' to detect whether there is any statistically significant difference between these means. A dependent *t* test is an example of a *within-subjects* or *repeated-measures* statistical test and is most appropriate for a research design in which the same subjects are tested more than once. Thus, in the dependent *t* test, 'related groups' indicate that the same subjects are tested twice on the same variable and the two sets of scores are recorded as the 'related groups' scores. In other words, the same subjects in one group have been measured on two occasions on the same dependent variable. For example, you might have measured 10 individuals' level of anxiety (the dependent variable) before and after they underwent a new form of therapy. You would like to know whether the therapy lowered their level of anxiety. For this scenario, you can use a dependent *t* test because you have two related groups (i.e., two sets of anxiety data generated by the same group of subjects pre- and post-therapy). The first related group consists of the subjects prior to the therapy and the second related group consists of the same subjects after receiving the therapy.

13.2 Independent *t* test

Let's use an example to demonstrate how the independent *t* test works. Let's say a researcher wants to investigate whether first-year male and female students at a university differ in their memory abilities. Ten male students and 10 female students were randomly selected from the first-year enrolment roll to serve as subjects. All 20 subjects were read 30 unrelated words and were then asked to recall as many of the words as possible. The numbers of words correctly recalled by each subject were recorded (Table 13.1).

TABLE 13.1

Number of Words Recalled by Males and Females

	Males (X)		Females (Y)
s1	16	s1	24
s2	14	s2	23
s3	18	s3	26
s4	25	s4	17
s5	17	s5	18
s6	14	s6	20
s7	19	s7	23
s8	21	s8	26
s9	16	s9	24
s10	17	s10	20

At the outset, the researcher should state the null hypothesis to be tested. Since the researcher is interested in investigating gender difference in memory abilities, a suitable null hypothesis would be *'there is no gender difference in number of words recalled'*. The independent *t* test will tell us whether the data are consistent with this or depart significantly from this expectation. (NB: recall that the null hypothesis is simply something to test against. The researcher might well expect a difference between males and females, but it would be difficult to predict the direction of that difference – Are males better? Are females better? Therefore, it is sensible to set a null hypothesis of 'no difference' and then to see whether the data depart from this.)

Since this problem involves two independent groups of subjects (males and females) and one dependent variable (words recalled), the independent *t* test is appropriate. The equation for the independent *t* test is

$$t_{obt} = \frac{M_x - M_y}{\sqrt{\dfrac{(\Sigma X^2 - (\Sigma X)^2/N_x) + (\Sigma Y^2 - (\Sigma Y)^2/N_y) * \left(\dfrac{1}{N_x} + \dfrac{1}{N_y}\right)}{N_x + N_y - 2}}}$$

where

M_x = Mean score for male subjects
M_y = Mean score for female subjects
ΣX^2 = Sum of the squared X (male) scores
$(\Sigma X)^2$ = Sum of the X (male) scores squared
ΣY^2 = Sum of the squared X (female) scores
$(\Sigma Y)^2$ = Sum of the Y (female) scores squared
N_x = Number of male subjects
N_y = Number of female subjects

Step 1: Calculate the appropriate statistic using the derived data in Table 13.2

$$t_{obt} = \frac{M_x - M_y}{\sqrt{\dfrac{(\Sigma X^2 - (\Sigma X)^2/N_x) + (\Sigma Y^2 - (\Sigma Y)^2/N_y) * \left(\dfrac{1}{N_x} + \dfrac{1}{N_y}\right)}{N_x + N_y - 2}}}$$

$$t_{obt} = \frac{17.7 - 22.1}{\sqrt{\dfrac{(3233 - (177)^2/10) + (4975 - (221)^2/10) * \left(\dfrac{1}{10} + \dfrac{1}{10}\right)}{10 + 10 - 2}}}$$

TABLE 13.2

Summed Words Recalled for Males and Females

	Males (X)	X²		Females (Y)	Y²
s1	16	256	s1	24	576
s2	14	196	s2	23	529
s3	18	324	s3	26	676
s4	25	625	s4	17	289
s5	17	289	s5	18	324
s6	14	196	s6	20	400
s7	19	361	s7	23	529
s8	21	441	s8	26	676
s9	16	256	s9	24	576
s10	17	289	s10	20	400
	$\Sigma X = 177$	$\Sigma X^2 = 3233$		$\Sigma Y = 221$	$\Sigma Y^2 = 4975$
	$M_x = 17.7$			$M_y = 22.1$	

$$t_{obt} = \frac{-4.4}{\sqrt{\dfrac{(100.1)+(90.9)*(0.2)}{18}}}$$

$$t_{obt} = \frac{-4.4}{\sqrt{2.1222}}$$

$$t_{obt} = \frac{-4.4}{1.4567}$$

$$t_{obt} = -3.02$$

Step 2: Evaluate the statistic

The sign of the computed t_{obt} value can be either positive or negative. The t_{obt} value will be positive if the first mean (M_x) is larger than the second (M_y) and negative if it is smaller. Once the t_{obt} value has been computed, the researcher has to compare this value with the tabulated t_{crit} value listed in the t-table of significance (see Table B in the Appendix). The t-table contains probabilities based on the t distribution. Essentially, this table of significance will inform the researcher as to whether the computed t_{obt} value is large enough to conclude that the difference between the groups is not likely to have been a chance finding, that is, the finding is significant.

To test the significance of the t_{obt} value, the researcher needs to do two things. First, he needs to set a probability level (called the alpha (α) level) for rejecting the null hypothesis. Let's say he sets the alpha level at $p < 0.05$. This means that if the difference between the two groups' means occurring by chance is equal to or less than five times out of a hundred, the researcher can reject the null hypothesis and conclude that there is a statistically significant difference between the means.

Second, in evaluating the computed t_{obt} value, the researcher also needs to determine the 'degrees of freedom' (*df*) for the test.

Degrees of freedom can be described as the number of scores that are free to vary. For example, suppose you tossed three dice. The total score adds up to 15. If you rolled a 6 on the first die and a 4 on the second, then the third die must be a 5 (otherwise, the total would not add up to 15). In this example, 2 dice are free to vary while the third is not. Therefore, there are 2 degrees of freedom.

For the independent *t* test, the 'degrees of freedom' is the sum of the participants in both groups minus 2. For this example, the *df* is $(n_1 + n_2) - 2 = 18$.

Given the alpha level (0.05), the *df* (18), and the t_{obt} value (−3.02), the researcher can look up the tabulated t_{crit} value in a standard *t* table of significance (see Table B in the Appendix) to determine whether the t_{obt} value is large enough to be significant. Specifically, if the calculated t_{obt} value exceeds the tabulated t_{crit} value, the researcher can reject the null hypothesis of no difference and conclude that the means are significantly different at the 0.05 probability level. That is,

$$\text{If } |t_{obt}| \geq |t_{crit}|, \text{ reject } H_o. \text{ If not, retain } H_o$$

For the present example, the *df* is 18. From Table B, with $\alpha = 0.05_{2\ tail}$, the tabulated $t_{crit} = \pm 2.101$. Since the computed t_{obt} value $= -3.02$ is bigger than the t_{crit} value $= \pm 2.101$, the researcher can reject the null hypothesis of no difference and conclude that there is a significant gender difference in the number of words recalled.

13.2.1 SPSS Windows Method: Independent *t* test

The data set has been saved under the name: **EX15.SAV**

1. From the menu bar, click **Analyze**, then **Compare Means**, and then **Independent-Samples *t* test.** The following window will open.

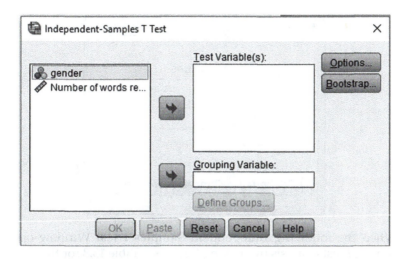

2. Since **GENDER** is the grouping (independent) variable, transfer it to the **Grouping Variable:** field by clicking (highlight) the variable and then clicking ⬥. As **WORDS** is the test (dependent) variable, transfer it to the **Test Variable(s):** field by clicking (highlight) the variable and then clicking ⬥.

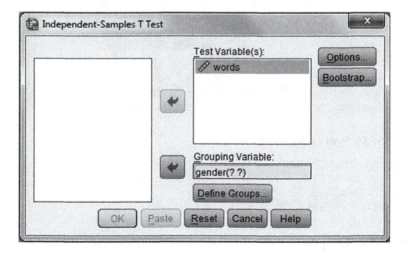

3. Click `Define Groups...` to define the range for the grouping variable of **GENDER** (coded 1 = male, 2 = female). When the following **Define Groups** Window opens, type **1** in the **Group 1:** field and **2** in the **Group 2:** field, and then click `Continue`.

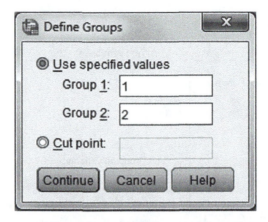

4. When the following **Independent-Samples *t* test** Window opens, run the *t* test analysis by clicking [OK]. See Table 13.3 for the results.

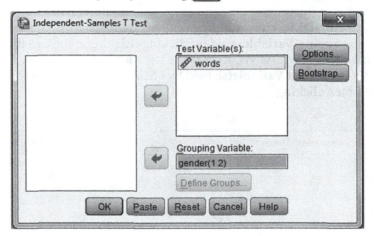

13.2.2 SPSS Syntax Method

```
T-TEST GROUPS=GENDER(1 2)
/MISSING=ANALYSIS
/VARIABLES=WORDS
/CRITERIA=CI(.95).
```

13.2.3 SPSS Output

See Table 13.3.

13.2.4 Results and Interpretation

The **Group Statistics** table informs the researcher that there are 10 partici-pants in each group and that for the male participants, the mean number of

TABLE 13.3

Independent *t* test Output

Group Statistics					
	Gender	**N**	**Mean**	**Std. Deviation**	**Std. Error Mean**
Number of words recalled	Male	10	17.7000	3.33500	1.05462
	Female	10	22.1000	3.17805	1.00499

Independent Samples Test									
	Levene's Test for Equality of Variances		**t-test for Equality of Means**						
								95% Confidence Interval of the Difference	
	F	**Sig**	**t**	**df**	**Sig. (2-tailed)**	**Mean Difference**	**Std. Error Difference**	**Lower**	**Upper**
WORDS Equal variances assumed	0.87	0.772	−3.020	18	0.007	−4.4000	1.4568	−7.4606	−1.3394
Equal variances not assumed			−3.020	17.958	0.007	−4.4000	1.4568	−7.4611	−1.3389

words recalled is 17.7 ($S_D = 3.335$), and that for the female participants, the mean number of words recalled is 22.1 ($S_D = 3.178$).

Looking at the **Independent Samples Test** table, the **Levene's Test for Equality of Variances** tests the assumption of **Homogeneity of variance.** That is, the statistic tests the hypothesis that the two population variances (from which the male and female samples were drawn) are equal. In this example, the Levene statistic is $F = 0.087$ and the corresponding level of significance is large (i.e., $p > 0.05$) (see Table 13.3). Thus, the assumption of homogeneity of variance has not been violated, and the **Equal variances assumed** *t* test statistic can be used for evaluating the null hypothesis of equality of means. If the significance level of the Levene statistic is small (i.e., $p < 0.05$), the assumption that the population variances are equal is rejected and the **Equal variances not assumed** *t* test statistic should be used.

The results from the *t* test analysis indicate the following:

- There is a significant difference between the male and female samples in the number of words correctly recalled, $t(df = 18) = −3.02$, $p < 0.01$. The mean values indicate that not only is there significant gender difference in the number of words correctly recalled, but also that females correctly recalled more words ($M = 22.10$) than males ($M = 17.70$).

- The **95% Confidence Interval of the Difference** shows that the null hypothesis value (i.e., zero, no difference) does not fall within this interval (Lower = −7.4606, Upper = −1.3394). Therefore, the null hypothesis of equality of means can be rejected.

13.3 Dependent/Correlated t test

The dependent t test (also called the correlated t test, the paired t test, or paired-samples t test) compares the means of two related groups to detect whether there are any statistically significant differences between these means. As mentioned earlier, 'related groups' indicate that the same subjects are present in both groups with the same subjects tested more than once. The dependent t test is an example of a 'within-subjects' or 'repeated-measures' statistical test and indicates that the same subjects are tested more than once. A common experiment of this type involves the *before and after* design. The test can also be used for the **matched group** design in which pairs of subjects that are matched on one or more characteristics (e.g., IQ, grades, and so forth) serve in the two conditions. As the subjects in the groups are matched and not independently assigned, this design is also referred to as a **correlated groups** design.

Let's use an example to demonstrate how the dependent t test works. Let's say a researcher designed an experiment to test the effect of drug X on eating behavior. The amount of food eaten by a group of 10 rats in a one-week period, prior to ingesting drug X, was recorded. The rats were then given drug X, and the amount of food eaten in a one-week period was again recorded. Table 13.4 shows the amount of food in grams eaten during the "before" and "after" conditions.

TABLE 13.4

Amount of Food in Grams Eaten during "before" and "after" Drug X

	Food Eaten	
	Before Ingesting Drug X	**After Ingesting Drug X**
s1	100	60
s2	180	80
s3	160	110
s4	220	140
s5	140	100
s6	250	200
s7	170	100
s8	220	180
s9	120	140
s10	210	130

Once again, the researcher should state the null hypothesis to be tested. Since the researcher is interested in investigating the effect of drug X on eating behavior, a suitable null hypothesis would be *'there is no difference in the amount of food eaten before and after drug X was ingested'*. The dependent *t* test will tell us whether the data are consistent with this or depart significantly from this expectation.

Since this problem involves one group of subjects (10 rats) and one dependent variable (amount of feed eaten), the dependent *t* test is appropriate. The equation for the dependent *t* test is

$$t = \frac{\bar{X}_D}{S_D/\sqrt{n}}$$

where

$\bar{X}_D$ = Mean difference between the two sets of scores
S_D = Standard deviation of the difference scores
n = Number of subjects

Step 1: Calculate the appropriate statistic using the derived data in Table 13.5

$$t = \frac{\bar{X}_D}{S_D/\sqrt{n}}$$

Calculate S_D (standard deviation of the difference scores). The equation for calculating S_D is

$$S_D = \sqrt{\frac{\Sigma x^2 - (\Sigma x)^2/n}{n-1}}$$

TABLE 13.5

Amount of Food Eaten before and after Ingesting Drug X

| | Food Eaten | | | |
	Before Drug X	After Drug X	Difference (D)	D²
s1	100	60	40	1600
s2	180	80	100	10,000
s3	160	110	50	2500
s4	220	140	80	6400
s5	140	100	40	1600
s6	250	200	50	2500
s7	170	100	70	4900
s8	220	180	40	1600
s9	120	140	−20	400
s10	210	130	80	6400
			$\Sigma D = 530$	$\Sigma D^2 = 37,900$
			$\bar{X}_D = 53$	

$$S_D = \sqrt{\frac{37900 - (530)^2/10}{10-1}}$$

$$S_D = \sqrt{\frac{9810}{9}}$$

$$S_D = \sqrt{1090} = 33.015$$

Thus,

$$t = \frac{\bar{X}_D}{S_D/\sqrt{n}}$$

$$t_{obt} = \frac{53}{33.015/\sqrt{10}}$$

$$t_{obt} = \frac{53}{10.44}$$

$$t_{obt} = 5.076$$

Step 2: Evaluate the statistic

Just as with the independent t test, once the dependent t_{obt} value has been computed the researcher has to compare this value with the tabulated t_{crit} value listed in the t-table of significance (see Table B in the Appendix). This table of significance will inform the researcher as to whether the computed t_{obt} value is large enough to conclude that the difference between the two sets of scores is not likely to have been a chance finding, that is, the finding is significant.

To test the significance of the t_{obt} value, assume that the researcher has set the alpha (α) level for rejecting the null hypothesis at $p < 0.05$. This means that if the difference between the two means occurring by chance is equal to or less than five times out of a hundred, the researcher can reject the null hypothesis and conclude that there is a statistically significant difference between the means.

For the dependent t test, the 'degrees of freedom' is the number of pairs of scores minus 1. For this example, the df is $n - 1 = 9$.

Given the alpha level (0.05), the df (9), and the t_{obt} value (5.076), the researcher can look up the tabulated t_{crit} value in a standard t table of significance (see Table B) to determine whether the t_{obt} value is large enough to be significant. Specifically, if the calculated t_{obt} value exceeds the tabulated t_{crit} value, the researcher can reject the null hypothesis of no difference and conclude that the means of the two sets of scores are significantly different at the 0.05 probability level. For the present example, the *df* is 9. From Table B, with $\alpha = 0.05_{2\,tail}$, the tabulated $t_{crit} = \pm 2.262$. Since the computed t_{obt} value $= 5.076$ is bigger than the $t_{crit} = \pm 2.262$, the researcher can reject the null hypothesis of no difference and conclude that there is a significant difference in the amount of food eaten before and after drug *X* was ingested.

13.3.1 SPSS Windows Method: Dependent *t* test

The data set has been saved under the name: **EX16.SAV**

1. From the menu bar, click **Analyze**, then **Compare Means**, and then **Paired-Samples *t* test.** The following window will open.

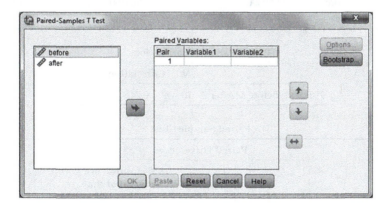

2. Transfer both the **BEFORE** and **AFTER** variables to the **Paired Variables:** field by clicking (highlight) these two variables, and then clicking ⬅. Click OK to run the *t*-test analysis. See Table 13.6 for the results.

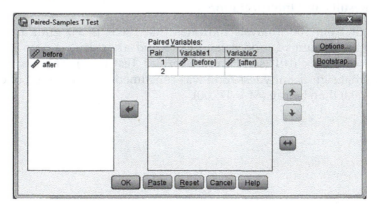

13.3.2 SPSS Syntax Method

```
T-TEST PAIRS=BEFORE WITH AFTER (PAIRED)
/CRITERIA=CI(.9500)
/MISSING=ANALYSIS.
```

13.3.3 SPSS Output

TABLE 13.6

Paired-Samples *t*-test Output

t-test					
Paired Samples Statistics					
		Mean	**N**	**Std. Deviation**	**Std. Error Mean**
Pair 1	Before	177.0000	10	48.31609	15.27889
	After	124.0000	10	43.25634	13.67886

		N	**Correlation**	**Sig.**
Paired Samples Correlations				
Pair 1	Before & After	10	0.745	0.013

Paired Samples Test

	Paired Differences							
				95% Confidence Interval of the Difference				
	Mean	**Std. Deviation**	**Std. Error Mean**	**Lower**	**Upper**	**t**	**df**	**Sign. (2-tailed)**
Pair 1 before – after	53.00000	33.01515	10.44031	29.38239	76.61761	5.076	9	0.001

13.3.4 Results and Interpretation

The result from the analysis indicates that there is a significant difference in the amount of food eaten before and after drug X was ingested, $t(df = 9) = 5.08$, $p < 0.01$ (see **Paired Samples Test** table). The mean values indicate that significantly less food was consumed after ingestion of drug X ($M = 124.00$) than before ($M = 177.00$).

14

Hypothesis Testing: One-Way Analysis of Variance

14.1 One-Way Analysis of Variance

The one-way analysis of variance (ANOVA) is an extension of the independent t test and is used to determine whether there are any significant differences between the means of three or more independent (unrelated) groups. What the one-way ANOVA does is to compare the means between the groups the researcher is interested in and determine whether any of those means are significantly different from each other. If the one-way ANOVA returns a significant result, the researcher can conclude that there are at least 2 group means that are significantly different from each other.

14.1.1 An Example

Let's use an example to demonstrate how the one-way ANOVA works. Let's say a researcher is interested in finding out whether intensity of electric shock will affect the time required to solve a set of difficult problems. Eighteen subjects were randomly assigned to the three experimental conditions of "low shock," "medium shock," and "high shock." The total time (in minutes) required to solve all the problems is the measure recorded for each subject (Table 14.1).

At the outset, the researcher should state the null hypothesis to be tested. Since the researcher is interested in investigating the effect of electric shock intensity on the time required to solve a set of difficult problems, a suitable null hypothesis would be *'there is no electric shock intensity difference in the time required to solve a set of difficult problems'*. The one-way ANOVA will inform the researcher whether the data are consistent with this or depart significantly from this expectation.

Since this problem involves three independent groups of subjects (low shock, medium shock, high shock) and one dependent variable (time taken to solve the problems), the one-way ANOVA is appropriate. In order to

TABLE 14.1

Total Time (in Minutes) Required to Solve Problems as a
Function of Three Levels of Electric Shock

	Shock Intensity	
Low	Medium	High
15	30	40
10	15	35
25	20	50
15	25	43
20	23	45
18	20	40

calculate the one-way ANOVA *F*-value, we need to compute the following parameters:

- SS_{total} (total sum of squares)
- $SS_{between}$ (between-groups sum of squares)
- SS_{within} (within-groups sum of squares)
- $MS_{between}$ (between-groups mean sum of squares)
- MS_{within} (within-groups mean sum of squares)
- $df_{between}$ (between-groups degrees of freedom)
- df_{within} (within-groups degrees of freedom)

Table 14.2 presents the summed values from Table 14.1.

$$SS_{total} = \left(\Sigma x_1^2 + \Sigma x_2^2 + \Sigma x_3^2\right) - \frac{(\Sigma x_1 + \Sigma x_2 + \Sigma x_3)^2}{n_1 + n_2 + n_3}$$

$$SS_{total} = (1899 + 3079 + 10799) - \frac{(103 + 133 + 253)^2}{18}$$

$$= 15777 - 13284.5$$

$$SS_{total} = \mathbf{2492.5}$$

$$SS_{between} = \left[\frac{(\Sigma x_1)^2}{n_1} + \frac{(\Sigma x_2)^2}{n_2} + \frac{(\Sigma x_3)^2}{n_3}\right] - \frac{(\Sigma x_1 + \Sigma x_2 + \Sigma x_3)^2}{n_1 + n_2 + n_3}$$

$$SS_{between} = \left[\frac{10609}{6} + \frac{17689}{6} + \frac{64009}{6}\right] - 13284.5$$

$$= 15384.5 - 13284.5$$

$$SS_{between} = \mathbf{2100}$$

TABLE 14.2

Summed Values for the Three Shock Intensity Groups

Low (x_1)	x_1^2	Medium (x_2)	x_2^2	High (x_3)	x_3^2
15	225	30	900	40	1600
10	100	15	225	35	1225
25	625	20	400	50	2500
15	225	25	625	43	1849
20	400	23	529	45	2025
18	324	20	400	40	1600
$\Sigma x_1 = 103$	$\Sigma x_1^2 = 1899$	$\Sigma x_2 = 133$	$\Sigma x_2^2 = 3079$	$\Sigma x_3 = 253$	$\Sigma x_3^2 = 10799$
$(\Sigma x_1)^2 = 10609$		$(\Sigma x_2)^2 = 17689$		$(\Sigma x_3)^2 = 64009$	
$M_1 = 17.17$		$M_2 = 22.17$		$M_3 = 42.17$	

$$SS_{within} = SS_{total} - SS_{between}$$

$$SS_{within} = 2492.5 - 2100 = \mathbf{392.5}$$

$$MS_{between} \text{(mean square between)}$$
$$= \frac{SS_{between}}{df \text{ between (number of groups subtract 1)}}$$

$$MS_{between} = \frac{2100}{2} = \mathbf{1050}$$

MS_{within}(mean square within)
$$= \frac{SS_{within}}{df \text{ within (number of subjects per group subtract } 1 * \text{number of groups)}}$$

$$MS_{within} = \frac{392.5}{15} = \mathbf{26.167}$$

$$F(df_{between}, df_{within}) = \frac{MS_{between}}{MS_{within}}$$

$$F(2,15) = \frac{1050}{26.17}$$

$$F(2,15) = \mathbf{40.12}$$

Source	SS	df	MS	F
Between	2100	2	1050	40.127
Within (error)	392.5	15	26.167	

The calculated F value is 40.12. In order to evaluate whether this value is statistically significant, the research must refer to the table of **critical values for the F distribution** (see Table C in the Appendix). To be significant, the F obtained value must be equal to or greater than the F critical value for the specified degrees of freedom. For this example, it can be seen that the relevant degrees of freedom are 2 (between-groups) and 15 (within-groups). Referring to Table C, go along 2 columns (df_x) and down 15 rows (df_y). The point of intersection yields the critical F value of 3.68. Since the obtained F value (40.12) is larger than this the researcher can conclude that the F value is statistically significant, $p < 0.05$. Based on this result, the researcher can reject the null hypothesis that the intensity of the electric shock has no effect on the time required to solve a set of difficult problems. That is, the overall difference in mean scores between the three shock-intensity groups is not due to chance.

At this point, it is important to note that the one-way ANOVA is an *omnibus* (overall) test statistic in that it will only inform the researcher whether there is an overall significant difference between the means of the groups. The test statistics cannot inform the researcher which specific groups are significantly different from each other, and only that at least two groups are. To determine which specific groups differed from each other, the researcher needs to conduct a follow-up *post hoc* test which pinpoints which group differs significantly from which.

14.1.2 Scheffé *Post Hoc* Test

Once the one-way ANOVA yields a significant omnibus F value, *post hoc* tests can be carried out to provide specific information on which means are significantly different from each other. There are a number of popular post hoc tests the researcher can choose from, such as the *least significant difference* (LSD) test, the *Tukey's* honest significant difference (HSD) test, and the *Scheffé* test. The Scheffé procedure is perhaps the most popular of the post hoc procedures, the most flexible, and the most conservative. It is also the most commonly applied post hoc procedure for follow-up multigroup comparisons. The Scheffé test computes an F statistic with $df = J - 1, N - J$, where

$$J = \text{number of groups (3)}$$
$$N = \text{Total number of subjects (18)}$$

TABLE 14.3

Matrix of Mean Score Differences between the Three Shock-Intensity Conditions

	Low Shock ($M = 17.17$)	Medium Shock ($M = 22.17$)	High Shock ($M = 42.17$)
Low shock	–	5.00	25.00**
Medium shock	–	–	20.00**

The equation for calculating the Scheffé test is as follows:

$$\sqrt{(J-1)(F_{critical})} * \sqrt{MS_{error}\left(\frac{1}{n_1} + \frac{1}{n_2}\right)}$$

For our example : $\sqrt{(2)(3.68)} * \sqrt{26.167 \, (0.1666 + 0.1666)}$

$$\sqrt{7.36} * \sqrt{8.687}$$

$$2.713 * 2.947 = \mathbf{7.996}$$

Thus, for a Scheffé F value of 7.996 with $df = 2$, 15, a mean difference between any two mean scores that is equal to or greater than 7.996 will be significant at the 0.01 level. Going back to our example, in order to see which mean scores generated from the three electric-shock intensity conditions are significantly different from which, set up a matrix of mean score differences between the three shock-intensity conditions as follows (Table 14.4).

Thus, the mean differences in Table 14.3 that are marked by asterisks exceed the Scheffé critical difference (7.996) and are significant at $p < 0.01$. This demonstrates that the High Shock condition is significantly different to both the Low and Medium Shock conditions that did not differ from each other.

14.1.3 SPSS Windows Method: One-Way ANOVA

The data set has been saved under the name: **EX17.SAV**

1. From the menu bar, click **Analyze**, then **Compare Means**, and then **One-Way ANOVA**. The following **One-Way ANOVA** Window will open.

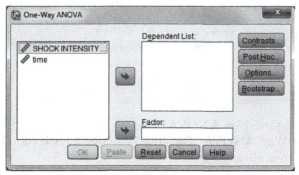

2. Transfer the dependent variable of **TIME** to the **Dependent List:** field by clicking (highlight) the variable and then clicking 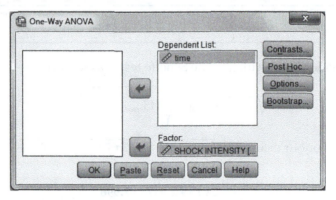. Transfer the independent variable of **SHOCK** to the **Factor:** field by clicking (highlight) the variable and then clicking ➡.

3. Since the one-way ANOVA will only perform an omnibus analysis of the *overall* differences between the three levels (low, medium, high) of the independent variable of **SHOCK**, it will not analyze the differences between the *specific* shock levels. To obtain multiple comparisons between the three shock levels (low shock vs. medium shock, low shock vs. high shock, medium shock vs. high shock) the researcher needs to perform a post hoc comparison test. Click Post Hoc... to achieve this. When the following **One-Way ANOVA: Post Hoc Multiple Comparisons** Window opens, check the **Scheffe** field to run the Scheffé post hoc test. Next, click Continue .

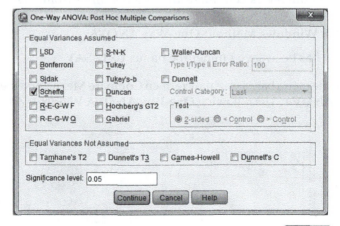

4. When the **One-Way ANOVA** Window opens, click Options... to open the **One-Way ANOVA: Options** Window. Check the **Descriptive** cell and then click Continue .

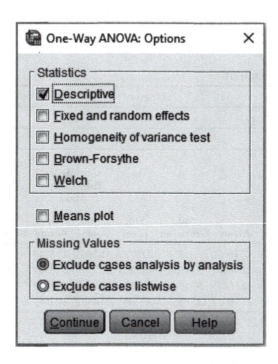

5. When the following **One-Way ANOVA** Window opens, run the analysis by clicking [OK]. See Table 14.4 for the results.

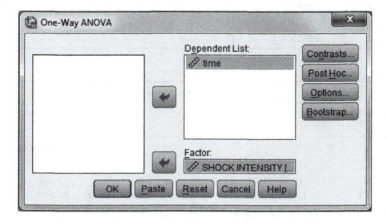

14.1.4 SPSS Syntax Method

```
ONEWAY TIME BY SHOCK
/STATISTICS DESCRIPTIVES
/MISSING ANALYSIS
/POSTHOC=SCHEFFE ALPHA(0.05).
```

14.1.5 SPSS Output

TABLE 14.4

One-Way ANOVA Output

	Oneway							
	Descriptives							
Time								
			Std.	Std.	95% Confidence Interval for Mean			
					Lower	Upper		
	N	Mean	Deviation	Error	Bound	Bound	Minimum	Maximum
LOW SHOCK	6	17.1667	5.11534	2.08833	11.7985	22.5349	10.00	25.00
MEDIUM SHOCK	6	22.1667	5.11534	2.08833	16.7985	27.5349	15.00	30.00
HIGH SHOCK	6	42.1667	5.11534	2.08833	36.7985	47.5349	35.00	50.00
Total	18	27.1667	12.10858	2.85402	21.1452	33.1881	10.00	50.00

	ANOVA				
Time					
	Sum of Squares	df	Mean Square	F	Sig.
Between groups	2100.000	2	1050.000	40.127	0.000
Within groups	392.500	15	26.167		
Total	2492.500	17			

	Post Hoc Tests					
	Multiple Comparisons					

Dependent Variable: TIME
Scheffe

(I) SHOCK INTENSITY	(J) SHOCK INTENSITY	Mean Difference $(I - J)$	Std. Error	Sig.	95% Confidence Interval	
					Lower Bound	Upper Bound
LOW SHOCK	MEDIUM SHOCK	−5.0000	2.95334	0.269	−13.0147	3.0147
	HIGH SHOCK	−25.0000[a]	2.95334	0.000	−33.0147	−16.9853
MEDIUM SHOCK	LOW SHOCK	5.0000	2.95334	0.269	−3.0147	13.0147
	HIGH SHOCK	−20.0000[a]	2.95334	0.000	−28.0147	−11.9853
HIGH SHOCK	LOW SHOCK	25.0000[a]	2.95334	0.000	16.9853	33.0147
	MEDIUM SHOCK	20.0000[a]	2.95334	0.000	11.9853	28.0147

[a] The mean difference is significant at the 0.05 level.

(Continued)

TABLE 14.4 (*Continued*)

One-Way ANOVA Output

	Homogeneous Subsets		
	Time		
SCHEFFE[a]			
		Subset for alpha = 0.05	
SHOCK INTENSITY	N	1	2
LOW SHOCK	6	17.1667	
MEDIUM SHOCK	6	22.1667	
HIGH SHOCK	6		42.1667
Sig.		0.269	1.000

Note: Means for groups in homogeneous subsets are displayed.
[a] Uses harmonic mean sample size = 6.000.

14.1.6 Results and Interpretation

The ANOVA table indicates that the intensity of the electric shock has a significant effect on the time taken to solve the problems, $F(2,15) = 40.12$, $p < 0.001$. The mean values for the three shock levels indicate that, as the shock level increased (from low to medium to high), so did the time taken to solve the problems (low: $M = 17.17$; medium: $M = 22.17$; high: $M = 42.17$).

14.1.7 Post Hoc Comparisons

While the highly significant *F*-value ($p < 0.001$) indicates that the means of the three shock levels differ significantly, it does not indicate the *location* of this difference. For example, the researcher may want to know whether the overall difference is due primarily to the difference between "low shock" and "high shock" levels, or between "low shock" and "medium shock" levels, or between "medium shock" and "high shock" levels. For this example, the Scheffé post hoc test was used to test for differences between specific shock levels.

In the **Multiple Comparisons** table, in the column labeled **Mean Difference (*I – J*)**, the mean difference values accompanied by asterisks indicate which shock levels differ significantly from each other at the 0.05 level of significance. The results indicate that the high shock level is significantly different from both the low shock and medium shock levels. The low shock level and the medium shock level do not differ significantly. These results show that the overall difference in time taken to solve complex problems between the three shock-intensity levels is due to the significantly greater amount of time taken by the subjects in the high shock condition.

15

Hypothesis Testing: Chi-Square Test

15.1 Nonparametric Tests

The *t* tests and the one-way ANOVA are termed *parametric* tests in that they depend considerably on population characteristics, or parameters, for their use. Both these tests, for instance, use the sample's mean and standard deviation statistics to estimate the values of the population parameters. Parametric tests also assume that the scores being analyzed come from populations that are *normally distributed* and have *equal variances*. In practice however, the data collected may violate one or both of these assumptions.

When the data collected flagrantly violate the above assumptions, the researcher must select an appropriate *nonparametric* test. Nonparametric inference tests have fewer requirements or assumptions about population characteristics. For example, to use these tests, it is not necessary to know the means, standard deviations, or shape of the population scores. Because nonparametric tests make no assumptions about the form of the populations from which the test samples were drawn, they are often referred to as *distribution-free* tests. A very common and popular nonparametric test is the *chi-square* (χ^2) test.

15.2 Chi-Square (χ^2) Test

The *t* test and ANOVA are used to analyze variables that are continuous in nature. Examples of continuous variables are time, weight, and height. Take time as an example. Between measurements of say 1 and 2 s, there is a continuous/infinite range from 1.0001 to 1.9999 s. But some types of research deal with nominal/categorical data, where observations are grouped into several discrete, mutually exclusive categories, and where one counts the frequency of occurrence in each category. For example, the researcher may wish to record how many individuals fall into a particular category, such as male or female, blue eyes or brown eyes, and Australian or Chinese. These counts are essentially frequencies and are noncontinuous and therefore must be treated differently from continuous data. Often the appropriate test is chi-square (χ^2), which the researcher can use to test whether the number of individuals in different categories fit a null hypothesis (an expectation of some sort). What the chi-square test does is to compare the observed frequencies of categories to frequencies that would be expected if the null hypothesis was true. The chi-square statistic is calculated by comparing the observed values against the expected values for each of the categories and examining the differences between them.

The chi-square test is most commonly used in two similar but distinct circumstances:

- For estimating how closely an observed distribution matches an expected distribution – this is referred to as the **goodness-of-fit test**
- For estimating whether two random variables are independent

15.2.1 Chi-Square Goodness-of-Fit Test

Let's use an example to demonstrate how this test works. Suppose we are interested in determining whether there is a difference among 18-year-olds living in Bangkok, Thailand, in their preference for three different brands of cola. We decide to conduct an experiment in which we randomly sample 42 18-year-olds and let them taste the three different brands. The data entered in each cell of Table 15.1 are the number or frequency of persons appropriate to that cell. Thus, 12 individuals preferred Brand A; 13 people preferred Brand B; and 17 participants preferred Brand C. Can we conclude from these data that there is a difference in preference in the population?

To answer this question, we first set up the null hypothesis to be tested. If there is no difference in the subjects' preference for the three different brands of cola, the null hypothesis would state that there will be equal number of subjects in the three cola brand cells. So, we divide our total number of subjects (42) by 3 to get our 'expected' values – 14 subjects per cell. At this point it should be noted that an important assumption underlying the chi-square test is that *the expected frequency for each category should be at least 5.*

Second, we set up a table showing both the observed frequencies and the expected frequencies (when the null hypothesis is true) (Table 15.2).

At first glance, comparison between the observed and expected frequencies appears to present strong evidence that there is indeed difference in preference for the three brands of cola and that more people prefer Brand C

TABLE 15.1

Frequency of Subjects in the Three Cola Brand Cells

Brand A	Brand B	Brand C	Total
12	13	17	42

TABLE 15.2

Observed and Expected Frequencies of Subjects in the Three Cola Brand Cells

	Brand A	Brand B	Brand C	Total
Observed	12	13	17	42
Expected (if the H_o is true)	14	14	14	42

than the other two brands. However, it is possible that such differences could have occurred by chance. The chi-square statistic can be used to test the null hypothesis that the likelihood of the observed differences did indeed occur by chance. The key aspect of the chi-square test is the comparison of observed and expected frequencies. That is, how many occurrences of an event can be expected, and how many of that event were observed to have happened? With the data set presented in Table 15.2, we can calculate the chi-square statistic as follows:

$$\chi^2_{obt} = \sum \frac{(f_o - f_e)^2}{f_e}$$

where
f_o = observed frequency
f_e = expected frequency

To calculate χ^2_{obt}, we simply sum the values of $(f_o - f_e)^2 / f_e$ for each cell. Thus,

$$\chi^2_{obt} = \sum \frac{(f_o - f_e)^2}{f_e}$$
$$= \frac{(12 - 14)^2}{14} + \frac{(13 - 14)^2}{14} + \frac{(17 - 14)^2}{14}$$
$$= 0.2857 + 0.0714 + 0.6428 = \mathbf{1.00}$$

In evaluating the statistical significance of the χ^2_{obt} value of 1.00, we need to first determine the degrees of freedom associated with the number of groups being compared. For our example, there are three groups, and therefore the observed frequencies of any two groups are free to vary. Once the frequencies of any two groups are known, the third group frequency score is fixed, since the sum of the three group frequencies must equal N (total number of subjects). For this example that involves just one variable, there are $k - 1$ degrees of freedom, where k equals the number of groups or categories. Thus, $df = 3 - 1 = 2$.

Once we have determined the degrees of freedom, we can compare our χ^2_{obt} value with the χ^2_{crit} value presented in Table D in the Appendix. As with the t test and the one-way ANOVA, the decision rule states that

If $\chi^2_{obt} \geq \chi^2_{crit}$, reject null hypothesis

Looking at Table D, we can see that for 2 degrees of freedom and $\alpha = 0.05$,

$$\chi^2_{crit} = 5.991$$

Since $\chi^2_{obt}(1.00) < \chi^2_{crit}(5.991)$, we fail to reject the null hypothesis, and conclude that there is no difference in the population regarding preference for the three brands of cola tested, that is, the differences in the observed and expected frequencies are due to chance.

15.2.2 SPSS Windows Method

The data set has been saved under the name: **EX18.SAV**

1. From the menu bar, click **Analyze, Nonparametric Tests, Legacy Dialogs**, and then **Chi-Square.** The following **Chi-Square Test** Window will open.

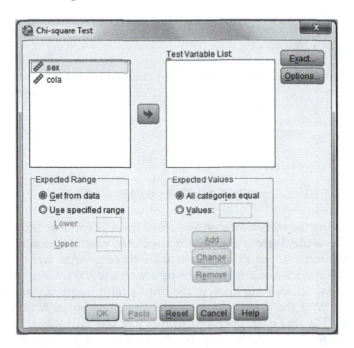

2. In the field containing the study's variables (**SEX, COLA**), click (highlight) the **COLA** variable, and then Click to transfer this variable to the **Test Variable List** field. Ensure that the **All categories equal** cell is checked (this is the default).

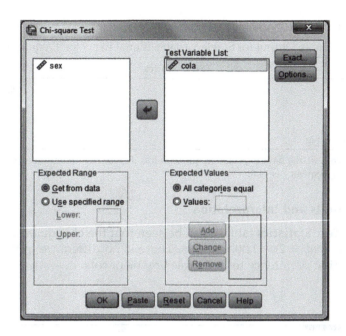

3. Click [OK] to run the chi-square analysis. See Table 15.3 for the results.

15.2.3 SPSS Syntax Method

```
NPAR TESTS CHISQUARE=COLA
/EXPECTED=EQUAL.
```

15.2.4 SPSS Output

TABLE 15.3

Chi-Square Output for Single Variable (Equal Expected Frequencies)

	NPar Tests	
	Chi-Square Test	
	Frequencies	
	Cola	
	Observed N	Expected N	Residual
Brand A	12	14.0	−2.0
Brand B	13	14.0	−1.0
Brand C	17	14.0	3.0
Total	42		

(Continued)

TABLE 15.3 (*Continued*)

Chi-Square Output for Single Variable (Equal Expected Frequencies)

Test Statistics

	Cola
Chi-square	1.000^a
df	2
Asymp. sig.	0.607

[a] 0 cells (0.0%) have expected frequencies less than 5. The minimum expected cell frequency is 14.0.

15.2.5 Results and Interpretation

From the **Test Statistics** table, it can be seen that the chi-square value is not significant, $\chi^2(df = 2) = 1.00$, $p > 0.05$. There is no difference in the population regarding preference for the three brands of cola.

15.3 Chi-Square (χ^2) Test of Independence between Two Variables

As stated earlier in this chapter, the other primary use of the chi-square test is to examine whether two categorical variables are independent or related. Very often in social science research, investigators are interested in finding factors that are related, such as ethnicity and IQ, education and income, gender and voting behavior, country of residence and support for euthanasia. In such cases, the chi-square test can be used to assess whether two categorical variables are independent or related.

To illustrate, let's suppose that as an extension to the above example, we are also interested in determining whether there is a relationship between preference for the 3 brands of cola and the gender of the 18-year-old subjects. The observed frequencies are shown in the 2 × 3 contingency Table 15.4. This example will employ the data set **EX18.SAV**.

A contingency table is a two-way table for examining relationships between categorical variables that have been classified into mutually exclusive categories and where the entries in the cells are frequency counts. Mutually exclusive categories denote the fact that a data point entered into a specific category cannot be entered in any other category.

In order to test whether gender and preference for the different brands of cola is related or not, we need to first set up the null hypothesis. The null hypothesis states that there is no contingency (relationship) between these variables in the population. Put in another way, the null hypothesis states that, in the population, preference for the three brands of cola and gender

TABLE 15.4

Gender and Preference for Cola Brand

	Brand A	Brand B	Brand C	*Row Marginal*
Male	4	2	15	21
Female	8	11	2	21
Column marginal	12	13	17	$N = 42$

are independent. If indeed the null hypothesis is true, then both the male and female subjects in the population should have the same proportion of individuals who prefer Brand A, Brand B, and Brand C. It is clear, however, that the contingency table shows different frequencies for males and females in the three brand columns. The 2×3 chi-square test of independence can be used to test the null hypothesis that these differences in frequencies are due to random sampling from a population in which the proportion of males is equal to the proportion of females in each of the categories.

As with the previous example, we test the null hypothesis by calculating the χ^2_{obt} and comparing it with the χ^2_{crit}. Unfortunately though, with studies involving two categorical variables, the calculation of the expected frequencies for each category is not that straightforward. As mentioned earlier, the null hypothesis states that the proportion of males in each category is the same as the proportion of females. If we know what these proportions are, we can simply multiply the number of males and females by their respective proportions to find the expected frequency for each category. For example, if the null hypothesis is true, the proportion of males and females in the population who prefer Brand A equals 0.50. Therefore, to find the expected frequencies for the male and female categories that prefer Brand A, all we need to do is to multiply 0.50 by the number of males and females (respectively) in the sample. Thus, the expected frequencies for the 'male-Brand A' category and for the 'female-Brand A' category would each equal $0.50(21) = 10.5$. Unfortunately, we do not know what the male and female proportions are in each of the categories. The question then becomes '*how do we compute these proportions?*'

The answer is that we have to estimate these proportions from the sample frequencies. From Table 15.4, it can be seen that 12 subjects (4 males and 8 females) out of the total sample of 42 preferred Brand A, 13 subjects (2 males and 11 females) out of the total sample of 42 preferred Brand B, and 17 subjects (15 males and 2 females) out of the total sample of 42 preferred Brand C. We can now use these sample proportions as our estimates of the population proportions when the null hypothesis is true, that is, independence between gender and preference for brands of cola. Thus, our estimates of the population proportions (when the null hypothesis is true) are as follows:

- Estimated population proportion for Brand A $= \dfrac{\text{Number of subjects who prefer Brand A}}{\text{Total number of subjects}} = \dfrac{12}{42}$

- Estimated population proportion for Brand B $=\dfrac{\text{Number of subjects who prefer Brand B}}{\text{Total number of subjects}}=\dfrac{13}{42}$

- Estimated population proportion for Brand C $=\dfrac{\text{Number of subjects who prefer Brand C}}{\text{Total number of subjects}}=\dfrac{17}{42}$

We can now use these population proportion estimates to calculate the expected frequencies (f_e) for each category as follows:

- *For Male-Brand A category*

$$f_e = (\text{Estimated population proportion for Brand A})$$
$$\times (\text{Total number of males})$$
$$= \dfrac{12}{42}(21) = 6$$

- *For Male-Brand B category*

$$f_e = (\text{Estimated population proportion for Brand B})$$
$$\times (\text{Total number of males})$$
$$= \dfrac{13}{42}(21) = \mathbf{6.5}$$

- *For Male-Brand C category*

$$f_e = (\text{Estimated population proportion for Brand C})$$
$$\times (\text{Total number of males})$$
$$= \dfrac{17}{42}(21) = \mathbf{8.5}$$

- *For Female-Brand A category*

$$f_e = (\text{Estimated population proportion for Brand A})$$
$$\times (\text{Total number of females})$$
$$= \dfrac{12}{42}(21) = \mathbf{6}$$

- *For Female-Brand B category*

$$f_e = (\text{Estimated population proportion for Brand B})$$
$$\times (\text{Total number of females})$$
$$= \dfrac{13}{42}(21) = \mathbf{6.5}$$

• *For Female-Brand C category*

$$f_e = \text{(Estimated population proportion for Brand C)}$$
$$\times \text{(Total number of females)}$$
$$= \frac{17}{42}(21) = \mathbf{8.5}$$

These calculated expected frequencies (in brackets) are presented along-side the observed frequencies in Table 15.5.

Once we have calculated the expected frequencies for each category, the next step is to calculate the χ^2_{obt}. To do this, we will use the same equation as in the previous example.

$$\chi^2_{obt} = \sum \frac{(f_o - f_e)^2}{f_e}$$
$$= \frac{(4-6)^2}{6} + \frac{(2-6.5)^2}{6.5} + \frac{(15-8.5)^2}{8.5} + \frac{(8-6)^2}{6} + \frac{(11-6.5)^2}{6.5} + \frac{(2-8.5)^2}{8.5}$$
$$= 0.6666 + 3.1154 + 4.9706 + 0.6666 + 3.1154 + 4.9706 = \mathbf{17.5052}$$

In evaluating the statistical significance of the χ^2_{obt} value of 17.5052, we need to first determine the degrees of freedom associated with the number of groups that are free to vary. For this example involving a 2×3 contin-gency table, only two groups' scores are free to vary and all the remaining groups' scores are fixed. Perhaps the easiest way to calculate the df for con-tingency tables is to apply the following equation:

$$df = (r-1)(c-1)$$

where
$r = $ number of rows
$c = $ number of columns

TABLE 15.5

Observed and Expected (in Brackets) Frequencies for 2 (Gender) × 3 (Cola Brand) Contingency Table

	Brand A	Brand B	Brand C	Row Marginal
Male	4 (6)	2 (6.5)	15 (8.5)	21
Female	8 (6)	11 (6.5)	2 (8.5)	21
Column marginal	12	13	17	$N = 42$

Applying the equation, we obtain

$$df = (2-1)(3-1) = 2$$

Once we have determined the degrees of freedom, we can compare our χ^2_{obt} value with the χ^2_{crit} value presented in Table D in the Appendix. As stated earlier,

$$\text{If } \chi^2_{obt} > \chi^2_{crit}, \quad \text{reject null hypothesis}$$

Looking at Table D, we can see that for 2° of freedom and $\alpha = 0.05$,

$$\chi^2_{crit} = 5.991$$

Since $\chi^2_{obt} > (17.5052) > \chi^2_{crit}(5.991)$, we reject the null hypothesis of independence, and conclude that there is a relationship in the population regarding gender and preference for the three brands of cola tested.

15.3.1 SPSS Windows Method

1. From the menu bar, click **Analyze**, then **Descriptive Statistics**, and then **Crosstabs....** The following **Crosstabs** Window will open.

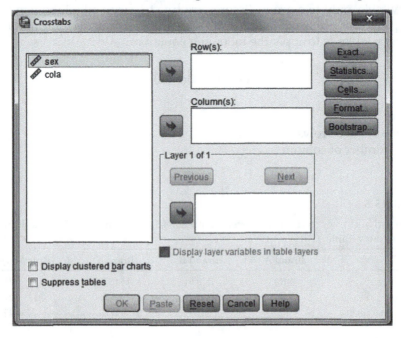

2. In the field containing the study's variables, click (highlight) the **COLA** variable, and then click 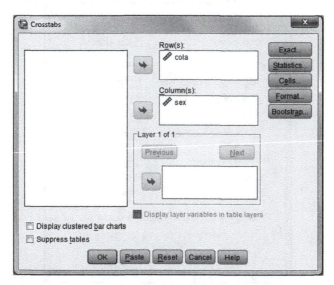 to transfer this variable to the **Row(s):** field. Next, click (highlight) the **SEX** variable, and then click to transfer this variable to the **Column(s):** field.

3. Click 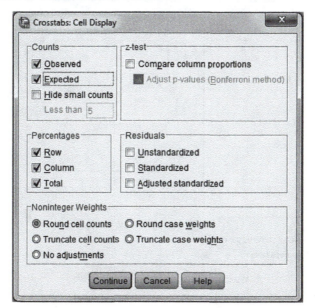 to open the **Crosstabs: Cell Display** Window below. Under **Counts**, check the **Observed** and **Expected** cells. Under **Percentages**, check the **Row, Column,** and **Total** cells. Click to return to the **Crosstabs** Window.

4. When the **Crosstabs** Window opens, click [Statistics] to open the
Crosstabs: Statistics Window below. Check the **Chi-square** cell,
and then click [Continue] to return to the Crosstabs Window.

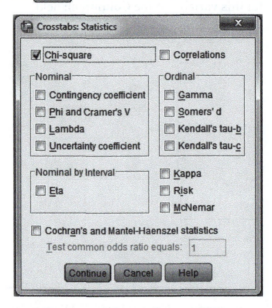

5. When the **Crosstabs** Window opens, click [OK] to run the analysis.
See Table 15.6 for the results.

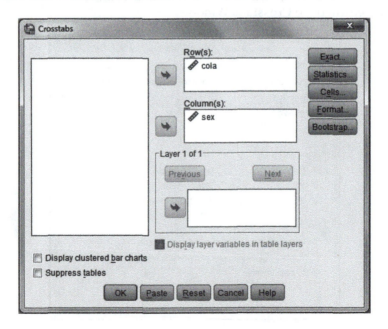

15.3.2 SPSS Syntax Method

```
CROSSTABS TABLES=COLA BY SEX
/CELLS=COUNT ROW COLUMN TOTAL EXPECTED
/STATISTICS=CHISQ.
```

15.3.3 SPSS Output

TABLE 15.6

Chi-Square Output for Test of Independence between Two Variables

			Crosstabs			
			Case Processing Summary			
				Cases		
		Valid		**Missing**		**Total**
		N	Percent	N	Percent	N	Percent
Cola[a] sex		42	100.0	0	0.0	42	100.0

		Cola[a] Sex Crosstabulation			
				Sex	
			Male	**Female**	**Total**
Cola	Brand A	Count	4	8	12
		Expected count	6.0	6.0	12.0
		% within cola	33.3	66.7	100.0
		% within sex	19.0	38.1	28.6
		% of total	9.5	19.0	28.6
	Brand B	Count	2	11	13
		Expected count	6.5	6.5	13.0
		% within cola	15.4	84.6	100.0
		% within sex	9.5	52.4	31.0
		% of total	4.8	26.2	31.0
	Brand C	Count	15	2	17
		Expected count	8.5	8.5	17.0
		% within cola	88.2	11.8	100.0
		% within sex	71.4	9.5	40.5
		% of total	35.7	4.8	40.5
Total		Count	21	21	42
		Expected count	21.0	21.0	42.0
		% within cola	50.0	50.0	100.0
		% within sex	100.0	100.0	100.0
		% of total	50.0	50.0	100.0

(Continued)

TABLE 15.6 (*Continued*)

Chi-Square Output for Test of Independence between Two Variables

	Chi-Square Tests		
	Value	*df*	Asymptotic Significance (2-sided)
Pearson chi-square	17.505[a]	2	0.000
Likelihood ratio	19.470	2	0.000
Linear-by-linear association	9.932	1	0.002
N of valid cases	42		

[a] 0 cells (0.0%) have expected count less than 5. The minimum expected count is 6.00.

15.3.4 Results and Interpretation

The results show that the **Expected Count** frequency in each of the 6 cells generated by the factorial combination of SEX and COLA is greater than 5. This means that the analysis has not violated a main assumption underlying the chi-square test.

The **Pearson Chi-Square** statistic is used to determine whether there is a significant relationship between preference for the 3 brands of cola and the gender of the 18-year-old subjects. The **Pearson Chi-Square** value is statistically significant, $\chi^2(df = 2) = 17.51$, $p < 0.001$. This means that preference for the three brands of cola varied as a function of the subject's gender. Looking at the **Cola*Sex Crosstabulation** table, it can be seen that the majority of the male subjects prefer Brand C (**Count** = 15; **% Within Sex** = 71.4%) over Brand A (**Count** = 4; **% Within Sex** = 19%) and Brand B (**Count** = 2; **% Within Sex** = 9.5%). For female subjects, their preference was for Brand B (**Count** = 11; **% Within Sex** = 52.4%), followed by Brand A (**Count** = 8; **% Within Sex** = 38.1%) and Brand C (**Count** = 2; **% Within Sex** = 9.5%).

Appendix

TABLE A

Z Scores: Areas under the Normal Curve

Standard Normal (Z) Table Area between 0 and z

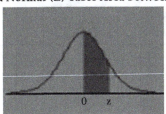

	0.00	0.01	0.02	0.03	0.04	0.05	0.06	0.07	0.08	0.09
0.0	0.0000	0.0040	0.0080	0.0120	0.0160	0.0199	0.0239	0.0279	0.0319	0.0359
0.1	0.0398	0.0438	0.0478	0.0517	0.0557	0.0596	0.0636	0.0675	0.0714	0.0753
0.2	0.0793	0.0832	0.0871	0.0910	0.0948	0.0987	0.1026	0.1064	0.1103	0.1141
0.3	0.1179	0.1217	0.1255	0.1293	0.1331	0.1368	0.1406	0.1443	0.1480	0.1517
0.4	0.1554	0.1591	0.1628	0.1664	0.1700	0.1736	0.1772	0.1808	0.1844	0.1879
0.5	0.1915	0.1950	0.1985	0.2019	0.2054	0.2088	0.2123	0.2157	0.2190	0.2224
0.6	0.2257	0.2291	0.2324	0.2357	0.2389	0.2422	0.2454	0.2486	0.2517	0.2549
0.7	0.2580	0.2611	0.2642	0.2673	0.2704	0.2734	0.2764	0.2794	0.2823	0.2852
0.8	0.2881	0.2910	0.2939	0.2967	0.2995	0.3023	0.3051	0.3078	0.3106	0.3133
0.9	0.3159	0.3186	0.3212	0.3238	0.3264	0.3289	0.3315	0.3340	0.3365	0.3389
1.0	0.3413	0.3438	0.3461	0.3485	0.3508	0.3531	0.3554	0.3577	0.3599	0.3621
1.1	0.3643	0.3665	0.3686	0.3708	0.3729	0.3749	0.3770	0.3790	0.3810	0.3830
1.2	0.3849	0.3869	0.3888	0.3907	0.3925	0.3944	0.3962	0.3980	0.3997	0.4015
1.3	0.4032	0.4049	0.4066	0.4082	0.4099	0.4115	0.4131	0.4147	0.4162	0.4177
1.4	0.4192	0.4207	0.4222	0.4236	0.4251	0.4265	0.4279	0.4292	0.4306	0.4319
1.5	0.4332	0.4345	0.4357	0.4370	0.4382	0.4394	0.4406	0.4418	0.4429	0.4441
1.6	0.4452	0.4463	0.4474	0.4484	0.4495	0.4505	0.4515	0.4525	0.4535	0.4545
1.7	0.4554	0.4564	0.4573	0.4582	0.4591	0.4599	0.4608	0.4616	0.4625	0.4633
1.8	0.4641	0.4649	0.4656	0.4664	0.4671	0.4678	0.4686	0.4693	0.4699	0.4706
1.9	0.4713	0.4719	0.4726	0.4732	0.4738	0.4744	0.4750	0.4756	0.4761	0.4767
2.0	0.4772	0.4778	0.4783	0.4788	0.4793	0.4798	0.4803	0.4808	0.4812	0.4817
2.1	0.4821	0.4826	0.4830	0.4834	0.4838	0.4842	0.4846	0.4850	0.4854	0.4857
2.2	0.4861	0.4864	0.4868	0.4871	0.4875	0.4878	0.4881	0.4884	0.4887	0.4890
2.3	0.4893	0.4896	0.4898	0.4901	0.4904	0.4906	0.4909	0.4911	0.4913	0.4916
2.4	0.4918	0.4920	0.4922	0.4925	0.4927	0.4929	0.4931	0.4932	0.4934	0.4936
2.5	0.4938	0.4940	0.4941	0.4943	0.4945	0.4946	0.4948	0.4949	0.4951	0.4952
2.6	0.4953	0.4955	0.4956	0.4957	0.4959	0.4960	0.4961	0.4962	0.4963	0.4964

(Continued)

TABLE A (*Continued*)

Z Scores: Areas under the Normal Curve

Standard Normal (Z) Table Area between 0 and z

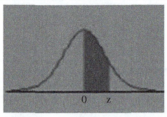

	0.00	0.01	0.02	0.03	0.04	0.05	0.06	0.07	0.08	0.09
2.7	0.4965	0.4966	0.4967	0.4968	0.4969	0.4970	0.4971	0.4972	0.4973	0.4974
2.8	0.4974	0.4975	0.4976	0.4977	0.4977	0.4978	0.4979	0.4979	0.4980	0.4981
2.9	0.4981	0.4982	0.4982	0.4983	0.4984	0.4984	0.4985	0.4985	0.4986	0.4986
3.0	0.4987	0.4987	0.4987	0.4988	0.4988	0.4989	0.4989	0.4989	0.4990	0.4990

TABLE B

Critical Values of *t* Distribution

1-tail $\alpha =$	0.1	0.05	0.025	0.01	0.005
2-tail $\alpha =$	0.2	0.1	0.05	0.02	0.01
$df = 1$	3.078	6.314	12.706	31.821	63.656
2	1.886	2.920	4.303	6.965	9.925
3	1.638	2.353	3.182	4.541	5.841
4	1.533	2.132	2.776	3.747	4.604
5	1.476	2.015	2.571	3.365	4.032
6	1.440	1.943	2.447	3.143	3.707
7	1.415	1.895	2.365	2.998	3.499
8	1.397	1.860	2.306	2.896	3.355
9	1.383	1.833	2.262	2.821	3.250
10	1.372	1.812	2.228	2.764	3.169
11	1.363	1.796	2.201	2.718	3.106
12	1.356	1.782	2.179	2.681	3.055
13	1.350	1.771	2.160	2.650	3.012
14	1.345	1.761	2.145	2.624	2.977
15	1.341	1.753	2.131	2.602	2.947
16	1.337	1.746	2.120	2.583	2.921
17	1.333	1.740	2.110	2.567	2.898
18	1.330	1.734	2.101	2.552	2.878
19	1.328	1.729	2.093	2.539	2.861
20	1.325	1.725	2.086	2.528	2.845
21	1.323	1.721	2.080	2.518	2.831

(*Continued*)

TABLE B (Continued)

Critical Values of *t* Distribution

22	1.321	1.717	2.074	2.508	2.819
23	1.319	1.714	2.069	2.500	2.807
24	1.318	1.711	2.064	2.492	2.797
25	1.316	1.708	2.060	2.485	2.787
26	1.315	1.706	2.056	2.479	2.779
27	1.314	1.703	2.052	2.473	2.771
28	1.313	1.701	2.048	2.467	2.763
29	1.311	1.699	2.045	2.462	2.756
30	1.310	1.697	2.042	2.457	2.750
60	1.296	1.671	2.000	2.390	2.660
120	1.289	1.658	1.980	2.358	2.617
∞	1.282	1.645	1.960	2.326	2.576

Note: The values listed in the table are the critical values of *t* for the specified degrees of freedom (left column) and the alpha level. To be significant, $|t_{obt}| \geq |t_{crit}|$

TABLE C

Critical Values for the *F* Distribution (for Use with ANOVA)

	Critical Values of *F* for the 0.05 Significance Level									
					df_x					
df_y	1	2	3	4	5	6	7	8	9	10
1	161.45	199.50	215.71	224.58	230.16	233.99	236.77	238.88	240.54	241.88
2	18.51	19.00	19.16	19.25	19.30	19.33	19.35	19.37	19.39	19.40
3	10.13	9.55	9.28	9.12	9.01	8.94	8.89	8.85	8.81	8.79
4	7.71	6.94	6.59	6.39	6.26	6.16	6.09	6.04	6.00	5.96
5	6.61	5.79	5.41	5.19	5.05	4.95	4.88	4.82	4.77	4.74
6	5.99	5.14	4.76	4.53	4.39	4.28	4.21	4.15	4.10	4.06
7	5.59	4.74	4.35	4.12	3.97	3.87	3.79	3.73	3.68	3.64
8	5.32	4.46	4.07	3.84	3.69	3.58	3.50	3.44	3.39	3.35
9	5.12	4.26	3.86	3.63	3.48	3.37	3.29	3.23	3.18	3.14
10	4.97	4.10	3.71	3.48	3.33	3.22	3.14	3.07	3.02	2.98
11	4.84	3.98	3.59	3.36	3.20	3.10	3.01	2.95	2.90	2.85
12	4.75	3.89	3.49	3.26	3.11	3.00	2.91	2.85	2.80	2.75
13	4.67	3.81	3.41	3.18	3.03	2.92	2.83	2.77	2.71	2.67
14	4.60	3.74	3.34	3.11	2.96	2.85	2.76	2.70	2.65	2.60
15	4.54	3.68	3.29	3.06	2.90	2.79	2.71	2.64	2.59	2.54
16	4.49	3.63	3.24	3.01	2.85	2.74	2.66	2.59	2.54	2.49
17	4.45	3.59	3.20	2.97	2.81	2.70	2.61	2.55	2.49	2.45
18	4.41	3.56	3.16	2.93	2.77	2.66	2.58	2.51	2.46	2.41
19	4.38	3.52	3.13	2.90	2.74	2.63	2.54	2.48	2.42	2.38

(Continued)

TABLE C (*Continued*)

Critical Values for the *F* Distribution (for Use with ANOVA)

	Critical Values of *F* for the 0.05 Significance level									
	df_x									
df_y	1	2	3	4	5	6	7	8	9	10
20	4.35	3.49	3.10	2.87	2.71	2.60	2.51	2.45	2.39	2.35
21	4.33	3.47	3.07	2.84	2.69	2.57	2.49	2.42	2.37	2.32
22	4.30	3.44	3.05	2.82	2.66	2.55	2.46	2.40	2.34	2.30
23	4.28	3.42	3.03	2.80	2.64	2.53	2.44	2.38	2.32	2.28
24	4.26	3.40	3.01	2.78	2.62	2.51	2.42	2.36	2.30	2.26
25	4.24	3.39	2.99	2.76	2.60	2.49	2.41	2.34	2.28	2.24
26	4.23	3.37	2.98	2.74	2.59	2.47	2.39	2.32	2.27	2.22
27	4.21	3.35	2.96	2.73	2.57	2.46	2.37	2.31	2.25	2.20
28	4.20	3.34	2.95	2.71	2.56	2.45	2.36	2.29	2.24	2.19
29	4.18	3.33	2.93	2.70	2.55	2.43	2.35	2.28	2.22	2.18
30	4.17	3.32	2.92	2.69	2.53	2.42	2.33	2.27	2.21	2.17
31	4.16	3.31	2.91	2.68	2.52	2.41	2.32	2.26	2.20	2.15
32	4.15	3.30	2.90	2.67	2.51	2.40	2.31	2.24	2.19	2.14
33	4.14	3.29	2.89	2.66	2.50	2.39	2.30	2.24	2.18	2.13
34	4.13	3.28	2.88	2.65	2.49	2.38	2.29	2.23	2.17	2.12
35	4.12	3.27	2.87	2.64	2.49	2.37	2.29	2.22	2.16	2.11
36	4.11	3.26	2.87	2.63	2.48	2.36	2.28	2.21	2.15	2.11
37	4.11	3.25	2.86	2.63	2.47	2.36	2.27	2.20	2.15	2.10
38	4.10	3.25	2.85	2.62	2.46	2.35	2.26	2.19	2.14	2.09
39	4.09	3.24	2.85	2.61	2.46	2.34	2.26	2.19	2.13	2.08
40	4.09	3.23	2.84	2.61	2.45	2.34	2.25	2.18	2.12	2.08
41	4.08	3.23	2.83	2.60	2.44	2.33	2.24	2.17	2.12	2.07
42	4.07	3.22	2.83	2.59	2.44	2.32	2.24	2.17	2.11	2.07
43	4.07	3.21	2.82	2.59	2.43	2.32	2.23	2.16	2.11	2.06
44	4.06	3.21	2.82	2.58	2.43	2.31	2.23	2.16	2.10	2.05
45	4.06	3.20	2.81	2.58	2.42	2.31	2.22	2.15	2.10	2.05
46	4.05	3.20	2.81	2.57	2.42	2.30	2.22	2.15	2.09	2.04
47	4.05	3.20	2.80	2.57	2.41	2.30	2.21	2.14	2.09	2.04
48	4.04	3.19	2.80	2.57	2.41	2.30	2.21	2.14	2.08	2.04
49	4.04	3.19	2.79	2.56	2.40	2.29	2.20	2.13	2.08	2.03
50	4.03	3.18	2.79	2.56	2.40	2.29	2.20	2.13	2.07	2.03
51	4.03	3.18	2.79	2.55	2.40	2.28	2.20	2.13	2.07	2.02
52	4.03	3.18	2.78	2.55	2.39	2.28	2.19	2.12	2.07	2.02
53	4.02	3.17	2.78	2.55	2.39	2.28	2.19	2.12	2.06	2.02
54	4.02	3.17	2.78	2.54	2.39	2.27	2.19	2.12	2.06	2.01
55	4.02	3.17	2.77	2.54	2.38	2.27	2.18	2.11	2.06	2.01
56	4.01	3.16	2.77	2.54	2.38	2.27	2.18	2.11	2.05	2.01
57	4.01	3.16	2.77	2.53	2.38	2.26	2.18	2.11	2.05	2.00

(*Continued*)

TABLE C (*Continued*)

Critical Values for the F Distribution (for Use with ANOVA)

	Critical Values of F for the 0.05 Significance level									
	df_x									
df_y	1	2	3	4	5	6	7	8	9	10
58	4.01	3.16	2.76	2.53	2.37	2.26	2.17	2.10	2.05	2.00
59	4.00	3.15	2.76	2.53	2.37	2.26	2.17	2.10	2.04	2.00
60	4.00	3.15	2.76	2.53	2.37	2.25	2.17	2.10	2.04	1.99
61	4.00	3.15	2.76	2.52	2.37	2.25	2.16	2.09	2.04	1.99
62	4.00	3.15	2.75	2.52	2.36	2.25	2.16	2.09	2.04	1.99
63	3.99	3.14	2.75	2.52	2.36	2.25	2.16	2.09	2.03	1.99
64	3.99	3.14	2.75	2.52	2.36	2.24	2.16	2.09	2.03	1.98
65	3.99	3.14	2.75	2.51	2.36	2.24	2.15	2.08	2.03	1.98
66	3.99	3.14	2.74	2.51	2.35	2.24	2.15	2.08	2.03	1.98
67	3.98	3.13	2.74	2.51	2.35	2.24	2.15	2.08	2.02	1.98
68	3.98	3.13	2.74	2.51	2.35	2.24	2.15	2.08	2.02	1.97
69	3.98	3.13	2.74	2.51	2.35	2.23	2.15	2.08	2.02	1.97
70	3.98	3.13	2.74	2.50	2.35	2.23	2.14	2.07	2.02	1.97
71	3.98	3.13	2.73	2.50	2.34	2.23	2.14	2.07	2.02	1.97
72	3.97	3.12	2.73	2.50	2.34	2.23	2.14	2.07	2.01	1.97
73	3.97	3.12	2.73	2.50	2.34	2.23	2.14	2.07	2.01	1.96
74	3.97	3.12	2.73	2.50	2.34	2.22	2.14	2.07	2.01	1.96
75	3.97	3.12	2.73	2.49	2.34	2.22	2.13	2.06	2.01	1.96
76	3.97	3.12	2.73	2.49	2.34	2.22	2.13	2.06	2.01	1.96
77	3.97	3.12	2.72	2.49	2.33	2.22	2.13	2.06	2.00	1.96
78	3.96	3.11	2.72	2.49	2.33	2.22	2.13	2.06	2.00	1.95
79	3.96	3.11	2.72	2.49	2.33	2.22	2.13	2.06	2.00	1.95
80	3.96	3.11	2.72	2.49	2.33	2.21	2.13	2.06	2.00	1.95
81	3.96	3.11	2.72	2.48	2.33	2.21	2.13	2.06	2.00	1.95
82	3.96	3.11	2.72	2.48	2.33	2.21	2.12	2.05	2.00	1.95
83	3.96	3.11	2.72	2.48	2.32	2.21	2.12	2.05	2.00	1.95
84	3.96	3.11	2.71	2.48	2.32	2.21	2.12	2.05	1.99	1.95
85	3.95	3.10	2.71	2.48	2.32	2.21	2.12	2.05	1.99	1.94
86	3.95	3.10	2.71	2.48	2.32	2.21	2.12	2.05	1.99	1.94
87	3.95	3.10	2.71	2.48	2.32	2.21	2.12	2.05	1.99	1.94
88	3.95	3.10	2.71	2.48	2.32	2.20	2.12	2.05	1.99	1.94
89	3.95	3.10	2.71	2.47	2.32	2.20	2.11	2.04	1.99	1.94
90	3.95	3.10	2.71	2.47	2.32	2.20	2.11	2.04	1.99	1.94
91	3.95	3.10	2.71	2.47	2.32	2.20	2.11	2.04	1.98	1.94
92	3.95	3.10	2.70	2.47	2.31	2.20	2.11	2.04	1.98	1.94
93	3.94	3.09	2.70	2.47	2.31	2.20	2.11	2.04	1.98	1.93
94	3.94	3.09	2.70	2.47	2.31	2.20	2.11	2.04	1.98	1.93
95	3.94	3.09	2.70	2.47	2.31	2.20	2.11	2.04	1.98	1.93

(*Continued*)

TABLE C (*Continued*)

Critical Values for the *F* Distribution (for Use with ANOVA)

	Critical Values of *F* for the 0.05 Significance level									
	df_x									
df_y	1	2	3	4	5	6	7	8	9	10
96	3.94	3.09	2.70	2.47	2.31	2.20	2.11	2.04	1.98	1.93
97	3.94	3.09	2.70	2.47	2.31	2.19	2.11	2.04	1.98	1.93
98	3.94	3.09	2.70	2.47	2.31	2.19	2.10	2.03	1.98	1.93
99	3.94	3.09	2.70	2.46	2.31	2.19	2.10	2.03	1.98	1.93
100	3.94	3.09	2.70	2.46	2.31	2.19	2.10	2.03	1.98	1.93

	Critical Values of *F* for the 0.01 Significance Level									
	df_x									
df_y	1	2	3	4	5	6	7	8	9	10
1	4052.19	4999.52	5403.34	5624.62	5763.65	5858.97	5928.33	5981.10	6022.50	6055.85
2	98.50	99.00	99.17	99.25	99.30	99.33	99.36	99.37	99.39	99.40
3	34.12	30.82	29.46	28.71	28.24	27.91	27.67	27.49	27.35	27.23
4	21.20	18.00	16.69	15.98	15.52	15.21	14.98	14.80	14.66	14.55
5	16.26	13.27	12.06	11.39	10.97	10.67	10.46	10.29	10.16	10.05
6	13.75	10.93	9.78	9.15	8.75	8.47	8.26	8.10	7.98	7.87
7	12.25	9.55	8.45	7.85	7.46	7.19	6.99	6.84	6.72	6.62
8	11.26	8.65	7.59	7.01	6.63	6.37	6.18	6.03	5.91	5.81
9	10.56	8.02	6.99	6.42	6.06	5.80	5.61	5.47	5.35	5.26
10	10.04	7.56	6.55	5.99	5.64	5.39	5.20	5.06	4.94	4.85
11	9.65	7.21	6.22	5.67	5.32	5.07	4.89	4.74	4.63	4.54
12	9.33	6.93	5.95	5.41	5.06	4.82	4.64	4.50	4.39	4.30
13	9.07	6.70	5.74	5.21	4.86	4.62	4.44	4.30	4.19	4.10
14	8.86	6.52	5.56	5.04	4.70	4.46	4.28	4.14	4.03	3.94
15	8.68	6.36	5.42	4.89	4.56	4.32	4.14	4.00	3.90	3.81
16	8.53	6.23	5.29	4.77	4.44	4.20	4.03	3.89	3.78	3.69
17	8.40	6.11	5.19	4.67	4.34	4.10	3.93	3.79	3.68	3.59
18	8.29	6.01	5.09	4.58	4.25	4.02	3.84	3.71	3.60	3.51
19	8.19	5.93	5.01	4.50	4.17	3.94	3.77	3.63	3.52	3.43
20	8.10	5.85	4.94	4.43	4.10	3.87	3.70	3.56	3.46	3.37
21	8.02	5.78	4.87	4.37	4.04	3.81	3.64	3.51	3.40	3.31
22	7.95	5.72	4.82	4.31	3.99	3.76	3.59	3.45	3.35	3.26
23	7.88	5.66	4.77	4.26	3.94	3.71	3.54	3.41	3.30	3.21
24	7.82	5.61	4.72	4.22	3.90	3.67	3.50	3.36	3.26	3.17
25	7.77	5.57	4.68	4.18	3.86	3.63	3.46	3.32	3.22	3.13
26	7.72	5.53	4.64	4.14	3.82	3.59	3.42	3.29	3.18	3.09

(*Continued*)

TABLE C (*Continued*)

Critical Values for the *F* Distribution (for Use with ANOVA)

	Critical Values of *F* for the 0.01 Significance Level									
	df_x									
df_y	1	2	3	4	5	6	7	8	9	10
27	7.68	5.49	4.60	4.11	3.79	3.56	3.39	3.26	3.15	3.06
28	7.64	5.45	4.57	4.07	3.75	3.53	3.36	3.23	3.12	3.03
29	7.60	5.42	4.54	4.05	3.73	3.50	3.33	3.20	3.09	3.01
30	7.56	5.39	4.51	4.02	3.70	3.47	3.31	3.17	3.07	2.98
31	7.53	5.36	4.48	3.99	3.68	3.45	3.28	3.15	3.04	2.96
32	7.50	5.34	4.46	3.97	3.65	3.43	3.26	3.13	3.02	2.93
33	7.47	5.31	4.44	3.95	3.63	3.41	3.24	3.11	3.00	2.91
34	7.44	5.29	4.42	3.93	3.61	3.39	3.22	3.09	2.98	2.89
35	7.42	5.27	4.40	3.91	3.59	3.37	3.20	3.07	2.96	2.88
36	7.40	5.25	4.38	3.89	3.57	3.35	3.18	3.05	2.95	2.86
37	7.37	5.23	4.36	3.87	3.56	3.33	3.17	3.04	2.93	2.84
38	7.35	5.21	4.34	3.86	3.54	3.32	3.15	3.02	2.92	2.83
39	7.33	5.19	4.33	3.84	3.53	3.31	3.14	3.01	2.90	2.81
40	7.31	5.18	4.31	3.83	3.51	3.29	3.12	2.99	2.89	2.80
41	7.30	5.16	4.30	3.82	3.50	3.28	3.11	2.98	2.88	2.79
42	7.28	5.15	4.29	3.80	3.49	3.27	3.10	2.97	2.86	2.78
43	7.26	5.14	4.27	3.79	3.48	3.25	3.09	2.96	2.85	2.76
44	7.25	5.12	4.26	3.78	3.47	3.24	3.08	2.95	2.84	2.75
45	7.23	5.11	4.25	3.77	3.45	3.23	3.07	2.94	2.83	2.74
46	7.22	5.10	4.24	3.76	3.44	3.22	3.06	2.93	2.82	2.73
47	7.21	5.09	4.23	3.75	3.43	3.21	3.05	2.92	2.81	2.72
48	7.19	5.08	4.22	3.74	3.43	3.20	3.04	2.91	2.80	2.72
49	7.18	5.07	4.21	3.73	3.42	3.20	3.03	2.90	2.79	2.71
50	7.17	5.06	4.20	3.72	3.41	3.19	3.02	2.89	2.79	2.70
51	7.16	5.05	4.19	3.71	3.40	3.18	3.01	2.88	2.78	2.69
52	7.15	5.04	4.18	3.70	3.39	3.17	3.01	2.87	2.77	2.68
53	7.14	5.03	4.17	3.70	3.38	3.16	3.00	2.87	2.76	2.68
54	7.13	5.02	4.17	3.69	3.38	3.16	2.99	2.86	2.76	2.67
55	7.12	5.01	4.16	3.68	3.37	3.15	2.98	2.85	2.75	2.66
56	7.11	5.01	4.15	3.67	3.36	3.14	2.98	2.85	2.74	2.66
57	7.10	5.00	4.15	3.67	3.36	3.14	2.97	2.84	2.74	2.65
58	7.09	4.99	4.14	3.66	3.35	3.13	2.97	2.84	2.73	2.64
59	7.09	4.98	4.13	3.66	3.35	3.12	2.96	2.83	2.72	2.64
60	7.08	4.98	4.13	3.65	3.34	3.12	2.95	2.82	2.72	2.63
61	7.07	4.97	4.12	3.64	3.33	3.11	2.95	2.82	2.71	2.63

(*Continued*)

TABLE C (*Continued*)

Critical Values for the *F* Distribution (for Use with ANOVA)

	Critical Values of *F* for the 0.01 Significance Level									
	df_x									
df_y	1	2	3	4	5	6	7	8	9	10
62	7.06	4.97	4.11	3.64	3.33	3.11	2.94	2.81	2.71	2.62
63	7.06	4.96	4.11	3.63	3.32	3.10	2.94	2.81	2.70	2.62
64	7.05	4.95	4.10	3.63	3.32	3.10	2.93	2.80	2.70	2.61
65	7.04	4.95	4.10	3.62	3.31	3.09	2.93	2.80	2.69	2.61
66	7.04	4.94	4.09	3.62	3.31	3.09	2.92	2.79	2.69	2.60
67	7.03	4.94	4.09	3.61	3.30	3.08	2.92	2.79	2.68	2.60
68	7.02	4.93	4.08	3.61	3.30	3.08	2.91	2.79	2.68	2.59
69	7.02	4.93	4.08	3.60	3.30	3.08	2.91	2.78	2.68	2.59
70	7.01	4.92	4.07	3.60	3.29	3.07	2.91	2.78	2.67	2.59
71	7.01	4.92	4.07	3.60	3.29	3.07	2.90	2.77	2.67	2.58
72	7.00	4.91	4.07	3.59	3.28	3.06	2.90	2.77	2.66	2.58
73	7.00	4.91	4.06	3.59	3.28	3.06	2.90	2.77	2.66	2.57
74	6.99	4.90	4.06	3.58	3.28	3.06	2.89	2.76	2.66	2.57
75	6.99	4.90	4.05	3.58	3.27	3.05	2.89	2.76	2.65	2.57
76	6.98	4.90	4.05	3.58	3.27	3.05	2.88	2.76	2.65	2.56
77	6.98	4.89	4.05	3.57	3.27	3.05	2.88	2.75	2.65	2.56
78	6.97	4.89	4.04	3.57	3.26	3.04	2.88	2.75	2.64	2.56
79	6.97	4.88	4.04	3.57	3.26	3.04	2.87	2.75	2.64	2.55
80	6.96	4.88	4.04	3.56	3.26	3.04	2.87	2.74	2.64	2.55
81	6.96	4.88	4.03	3.56	3.25	3.03	2.87	2.74	2.63	2.55
82	6.95	4.87	4.03	3.56	3.25	3.03	2.87	2.74	2.63	2.55
83	6.95	4.87	4.03	3.55	3.25	3.03	2.86	2.73	2.63	2.54
84	6.95	4.87	4.02	3.55	3.24	3.03	2.86	2.73	2.63	2.54
85	6.94	4.86	4.02	3.55	3.24	3.02	2.86	2.73	2.62	2.54
86	6.94	4.86	4.02	3.55	3.24	3.02	2.85	2.73	2.62	2.53
87	6.94	4.86	4.02	3.54	3.24	3.02	2.85	2.72	2.62	2.53
88	6.93	4.86	4.01	3.54	3.23	3.01	2.85	2.72	2.62	2.53
89	6.93	4.85	4.01	3.54	3.23	3.01	2.85	2.72	2.61	2.53
90	6.93	4.85	4.01	3.54	3.23	3.01	2.85	2.72	2.61	2.52
91	6.92	4.85	4.00	3.53	3.23	3.01	2.84	2.71	2.61	2.52
92	6.92	4.84	4.00	3.53	3.22	3.00	2.84	2.71	2.61	2.52
93	6.92	4.84	4.00	3.53	3.22	3.00	2.84	2.71	2.60	2.52
94	6.91	4.84	4.00	3.53	3.22	3.00	2.84	2.71	2.60	2.52
95	6.91	4.84	4.00	3.52	3.22	3.00	2.83	2.70	2.60	2.51
96	6.91	4.83	3.99	3.52	3.21	3.00	2.83	2.70	2.60	2.51

(*Continued*)

TABLE C (*Continued*)

Critical Values for the *F* Distribution (for Use with ANOVA)

	Critical Values of *F* for the 0.01 Significance Level									
					df_x					
df_y	1	2	3	4	5	6	7	8	9	10
97	6.90	4.83	3.99	3.52	3.21	2.99	2.83	2.70	2.60	2.51
98	6.90	4.83	3.99	3.52	3.21	2.99	2.83	2.70	2.59	2.51
99	6.90	4.83	3.99	3.52	3.21	2.99	2.83	2.70	2.59	2.51
100	6.90	4.82	3.98	3.51	3.21	2.99	2.82	2.69	2.59	2.50

How to use this table:

There are two tables here. The first one gives critical values of *F* at the $p = 0.05$ level of significance. The second table gives critical values of *F* at the $p = 0.01$ level of significance.

1. Obtain your *F*-ratio. This has (x, y) degrees of freedom associated with it. df_x = Between groups degrees of freedom; df_y = Within groups degrees of freedom.
2. Go along *x* columns, and down *y* rows. The point of intersection is your critical *F*-ratio.
3. If your obtained value of *F* is equal to or larger than this critical *F*-value, then your result is significant at that level of probability.

   An example: I obtain an *F* ratio of 3.96 with (2, 24) degrees of freedom.

   I go along 2 columns and down 24 rows. The critical value of *F* is 3.40. My obtained *F*-ratio is larger than this, and so I conclude that my obtained *F*-ratio is likely to occur by chance with a $p < 0.05$.

TABLE D

Critical Values of the Chi Square Distribution

	Level of Significance								
df	0.200	0.100	0.075	0.050	0.025	0.010	0.005	0.001	0.0005
1	1.642	2.706	3.170	3.841	5.024	6.635	7.879	10.828	12.116
2	3.219	4.605	5.181	5.991	7.378	9.210	10.597	13.816	15.202
3	4.642	6.251	6.905	7.815	9.348	11.345	12.838	16.266	17.731
4	5.989	7.779	8.496	9.488	11.143	13.277	14.860	18.467	19.998
5	7.289	9.236	10.008	11.070	12.833	15.086	16.750	20.516	22.106
6	8.558	10.645	11.466	12.592	14.449	16.812	18.548	22.458	24.104
7	9.803	12.017	12.883	14.067	16.013	18.475	20.278	24.322	26.019
8	11.030	13.362	14.270	15.507	17.535	20.090	21.955	26.125	27.869
9	12.242	14.684	15.631	16.919	19.023	21.666	23.589	27.878	29.667
10	13.442	15.987	16.971	18.307	20.483	23.209	25.188	29.589	31.421
11	14.631	17.275	18.294	19.675	21.920	24.725	26.757	31.265	33.138
12	15.812	18.549	19.602	21.026	23.337	26.217	28.300	32.910	34.822
13	16.985	19.812	20.897	22.362	24.736	27.688	29.820	34.529	36.479
14	18.151	21.064	22.180	23.685	26.119	29.141	31.319	36.124	38.111
15	19.311	22.307	23.452	24.996	27.488	30.578	32.801	37.698	39.720

(*Continued*)

TABLE D (*Continued*)

Critical Values of the Chi Square Distribution

df	\multicolumn{9}{c}{Level of Significance}
	0.200	0.100	0.075	0.050	0.025	0.010	0.005	0.001	0.0005
16	20.465	23.542	24.716	26.296	28.845	32.000	34.267	39.253	41.309
17	21.615	24.769	25.970	27.587	30.191	33.409	35.719	40.791	42.881
18	22.760	25.989	27.218	28.869	31.526	34.805	37.157	42.314	44.435
19	23.900	27.204	28.458	30.144	32.852	36.191	38.582	43.821	45.974
20	25.038	28.412	29.692	31.410	34.170	37.566	39.997	45.315	47.501
21	26.171	29.615	30.920	32.671	35.479	38.932	41.401	46.798	49.013
22	27.301	30.813	32.142	33.924	36.781	40.289	42.796	48.269	50.512
23	28.429	32.007	33.360	35.172	38.076	41.639	44.182	49.729	52.002
24	29.553	33.196	34.572	36.415	39.364	42.980	45.559	51.180	53.480
25	30.675	34.382	35.780	37.653	40.646	44.314	46.928	52.620	54.950
26	31.795	35.563	36.984	38.885	41.923	45.642	48.290	54.053	56.409
27	32.912	36.741	38.184	40.113	43.195	46.963	49.645	55.477	57.860
28	34.027	37.916	39.380	41.337	44.461	48.278	50.994	56.894	59.302
29	35.139	39.087	40.573	42.557	45.722	49.588	52.336	58.302	60.738
30	36.250	40.256	41.762	43.773	46.979	50.892	53.672	59.704	62.164
40	47.269	51.805	53.501	55.759	59.342	63.691	66.766	73.403	76.097
50	58.164	63.167	65.030	67.505	71.420	76.154	79.490	86.662	89.564
60	68.972	74.397	76.411	79.082	83.298	88.380	91.952	99.609	102.698
70	79.715	85.527	87.680	90.531	95.023	100.425	104.215	112.319	115.582
80	90.405	96.578	98.861	101.880	106.629	112.329	116.321	124.842	128.267
90	101.054	107.565	109.969	113.145	118.136	124.117	128.300	137.211	140.789
100	111.667	118.498	121.017	124.342	129.561	135.807	140.170	149.452	153.174

TABLE E

Glossary of SPSS Syntax Files

Frequency Analysis (Chapter 2: TRIAL.SAV)

FREQUENCIES VARIABLES=ALL *or list of variables*
/STATISTICS= MEAN MEDIAN MODE STDDEV *or* ALL *for all descriptive statistics.*

Frequency Analysis (Chapter 4: EX1.SAV)

FREQUENCIES VARIABLES=IQ.
FREQUENCIES VARIABLES=GROUP.

Compute P_{50} (the 50th percentile) (Chapter 4: EX2.SAV)

COMPUTE P50=XL+((i/fi)*(cum_fP-cum_fL)).
EXECUTE.

Compute P_{80} (the 80th percentile) (Chapter 4: EX3.SAV)

COMPUTE P80=XL+((i/fi)*(cum_fP-cum_fL)).
EXECUTE.

(*Continued*)

TABLE E *(Continued)*

Glossary of SPSS Syntax Files

Compute PR_{127} (percentile rank of 127) (Chapter 4: EX4.SAV)
```
COMPUTE PR127=(cum_fL+((fi/i)*(X-XL)))/N*100.
EXECUTE.
```

Compute PR_{112} (percentile rank of 112) (Chapter 4: EX5.SAV)
```
COMPUTE PR112=(cum_fL+((fi/i)*(X-XL)))/N*100.
EXECUTE.
```

Draw Bar Graph (Chapter 5: EX6.SAV)
```
GRAPH
/BAR(SIMPLE)=COUNT BY EMPLOY.
```

Draw Histogram (Chapter 5: EX7.SAV)
```
GRAPH
/HISTOGRAM=GROUP.
```

Draw Frequency Polygon of Grouped Frequency Distribution (Chapter 5: EX7.SAV)
```
GRAPH
/LINE(SIMPLE)=COUNT BY GROUP.
```

Draw Cumulative Percentage Curve of Frequency Distribution (Chapter 5: EX7.SAV)
```
GRAPH
/LINE(SIMPLE)=CUPCT BY IQ.
```

Calculate the Arithmetic Mean for the Set of 100 IQ Scores Presented in Table 4.2 (Chapter 6: EX1.SAV)
```
FREQUENCIES VARIABLES=IQ
/STATISTICS=MEAN
/ORDER=ANALYSIS.
```

Calculate the Mean from Grouped Frequency Distribution (Chapter 6: EX8.SAV)
```
COMPUTE fx = f*x.
COMPUTE MEAN = (308+441+420+1197+1638+1785+2576+1260+784+637+420)/100.
EXECUTE.
```

Calculate the Overall Mean for the OILCOM Shares Example (Chapter 6: EX9.SAV)
```
COMPUTE OVERALL _ MEAN = ((N1*LOT1) + (N2*LOT2) + (N3*LOT3))/(N1 + N2 + N3).
EXECUTE.
```

Calculate the Mode from a Histogram Distribution (Chapter 6: EX10.SAV)
```
GRAPH
/HISTOGRAM=SCORES.
```

Calculate the Range, the Standard Deviation, and the Variance for the Set of 100 IQ Scores Presented in Table 4.2 (Chapter 7: EX1.SAV)
```
FREQUENCIES VARIABLES=IQ
/STATISTICS=STDDEV VARIANCE RANGE
/ORDER=ANALYSIS.
```

(Continued)

TABLE E (*Continued*)

Glossary of SPSS Syntax Files

Calculate the *Percentile Rank* of the Test Score of 85, i.e., the Percentage of Scores that is Lower than the Score of 85 (Chapter 8: EX11.SAV)

```
DESCRIPTIVES VARIABLES=TEST_SCORES
/SAVE
/STATISTICS=MEAN STDDEV MIN MAX.
```

Calculate the 90th *Percentile*, i.e., the Exam Score below Which 90% of the Class's Scores Will Fall (Chapter 8: EX11.SAV)

```
FREQUENCIES VARIABLES=TEST_SCORES
/FORMAT=NOTABLE
/PERCENTILES=90.0
/ORDER=ANALYSIS.
```

Calculate the Lower and Upper Bound Scores that Bound the Middle 70% of the Statistics Exam's Distribution (Chapter 8: EX11.SAV)

```
FREQUENCIES VARIABLES=TEST_SCORES
/FORMAT=NOTABLE
/PERCENTILES=85.0 15.0
/ORDER=ANALYSIS.
```

Simple Scatter Plot of Two Variables (Weight, Height) to Show Whether a Relationship Exists (Chapter 9: EX12.SAV)

```
GRAPH
/SCATTERPLOT(BIVAR)=HEIGHT WITH WEIGHT
/MISSING=LISTWISE.
```

Convert Raw Scores into z Scores (Chapter 9: EX12.SAV)

```
DESCRIPTIVES VARIABLES=WEIGHT HEIGHT
/SAVE
/STATISTICS=MEAN STDDEV MIN MAX.
```

Calculate the Relationship between Two Continuous Variables (GPA, Reading Score) (Chapter 9: EX13.SAV)

```
CORRELATIONS
/VARIABLES=READ GPA
/PRINT=TWOTAIL NOSIG
/MISSING=PAIRWISE.
```

Construct the Least-Squares Regression Line: Predicting *Y* (GPA) from *X* (READ) (Chapter 10: EX13.SAV)

```
REGRESSION VARIABLES=(COLLECT)
/MISSING LISTWISE
/STATISTICS=DEFAULTS CI
/DEPENDENT=GPA
/METHOD=ENTER READ.
```

Calculate the 95% Confidence Interval for the Population Mean (Chapter 11: EX14.SAV)

```
EXAMINE VARIABLES=WEIGHT
/STATISTICS DESCRIPTIVES
```

(*Continued*)

TABLE E (*Continued*)

Glossary of SPSS Syntax Files

```
/CINTERVAL 95
/MISSING LISTWISE
/NOTOTAL.
```

Independent *t* test: (Chapter 13: EX15.SAV)
```
T-TEST GROUPS=GENDER(1 2)
/MISSING=ANALYSIS
/VARIABLES=WORDS
/CRITERIA=CI(0.95).
```

Dependent *t* test: (Chapter 13: EX16.SAV)
```
T-TEST PAIRS=BEFORE WITH AFTER (PAIRED)
/CRITERIA=CI(0.9500)
/MISSING=ANALYSIS.
```

One-Way ANOVA with Post Hoc Scheffé Test (Chapter 14: EX17.SAV)
```
ONEWAY TIME BY SHOCK
/STATISTICS DESCRIPTIVES
/MISSING ANALYSIS
/POSTHOC=SCHEFFE ALPHA(0.05).
```

Chi-Square Goodness-of-Fit Test (Chapter 15: EX18.SAV)
```
NPAR TESTS CHISQUARE=COLA
/EXPECTED=EQUAL.
```

Chi-Square (χ^2) Test of Independence between Two Variables (Chapter 15: EX18.SAV)
```
CROSSTABS TABLES=COLA BY SEX
/CELLS=COUNT ROW COLUMN TOTAL EXPECTED
/STATISTICS=CHISQ.
```

Bibliography

Argyrous, G. 1996. *Statistics for Social Research.* Macmillan Education Australia Pty Ltd., South Melbourne, Australia.

Minium, E. W., Clarke, R. B., and Coladarci, T. 1998. *Elements of Statistical Reasoning* (2nd ed.). Wiley, New York.

Pagano, R. R. 2013. *Understanding Statistics in the Behavioral Sciences* (10th ed.). Wadsworth Cengage Learning, Belmont, CA.

Index

A

Addition rule, 182; *see also* Probability
  equation for, 182
  using multiplication and, 188–190
  for mutually exclusive events, 183
  for non-mutually exclusive, 184
  simplified equation, 183
Alternative hypotheses, 205; *see also*
  Hypothesis testing
Analysis of variance (ANOVA), 229;
  see also One-way analysis of
  variance
ANOVA, *see* Analysis of variance
Arithmetic mean, 86; *see also* Central
  tendency
  calculation, 87
  data entry format, 92, 96–97
  example, 91–92, 96
  from grouped frequency
  distribution, 91, 92
  mean, median, and mode,
  107–109
  overall mean, 94–96, 96
  properties of, 98–101
  SPSS output, 90–91, 93–94, 98
  SPSS syntax method, 89–90,
  92–93, 97
  SPSS Window method, 87–89

B

Bar graph, 61; *see also* Graphing
  data entry format, 62
  example, 61
  SPSS bar graph output, 66–67
  SPSS syntax method, 64–65
  SPSS Windows method, 63–64
Bell-shaped curve, *see* Normal
  distribution
Between-groups design, 2–3; *see also*
  Scientific methodology of
  research

C

Central tendency, 85; *see also* Arithmetic
  mean; Frequency distributions;
  Median; Mode
  measures of, 86
  symmetry and skewness, 109
Chance hypothesis, 206; *see also*
  Hypothesis testing
Chi square distribution, critical values
  of, 261–262
Chi-square test, 239; *see also* Hypothesis
  testing
  chi-square goodness-of-fit test,
  240–242
  contingency table, 244
  critical values of chi square
  distribution, 261–262
  of independence between two
  variables, 244
  mutually exclusive categories, 244
  null hypothesis, 244
  results and interpretation, 244, 252
  SPSS output, 243, 251–252
  SPSS syntax method, 243–244, 251
  SPSS Windows method, 242–243,
  248–250
Confidence interval, 196; *see also*
  Inferential statistics
  calculation, 198
  SPSS output, 202
  SPSS syntax method, 200–201
  SPSS Windows method, 198–200
Confidence level, 197; *see also* Confidence
  interval
Contingency table, 244
Continuous variables, 32; *see also*
  Probability; Variables
  computing probability for,
  190–192
  real limits of, 32
Cookbook method, 9; *see also* SPSS
Correlated *t* test, *see* Dependent *t* test